FLIGHT TO ANYWHERE

A Navigator's Life

Mark Farmer

To CARL + ANNIE

with best wishes

+ RESPECT

[signature] 4/27/01

Coralreef Group
Post Office Box 2299
Yountville, California 94599

Printed in the United States of America
ISBN 0-9664832-0-0

Cover designed by Alan Worley

Coralreef Group
P.O. Box 2299
Yountville, CA 94599

CONTENTS

List of Illustrations

This book is dedicated to the several generations of air navigators who served until they were replaced by satellites and computers.

Introduction

Because I have lived during the era covered by Mark Farmer's stories, I may feel a certain affinity for them. However, I believe that other generations will be equally attracted and will smile and chuckle as I did.

Mark doesn't try to give global history (except to the extent that his travels do cover the world). He gives a slice, a taste, a segment of what went on in his life during the momentous events in a significant part of our century.

It would be too bad if someone didn't try to capture what went on at the day-to-day "working level." We have a great deal on grand strategy and summit conferences, but probably not enough on those who were performing the hands-on execution.

Curtis Hooper O'Sullivan
Brigadier General, Cal ARNG-Ret
Soldier, Historian

Acknowledgments

I am indebted to Lee Carbone for his first-hand account of the ammunition depot explosion in Nouméa, New Caledonia.

To Stan Smith who pushed us out of the nest, but remained in the classroom to offer support. I am grateful.

To Eleanor Knowles Dugan for her editorial help.

To my daughter Dulce Farmer and Wade Wenzel who made themselves available when computer help was needed. Without them I would have had to read the computer manuals.

And, to my wife, Ellen, who continues what her father had started in New York by planning entertainment for corporate events worldwide. I am so glad she came down the hall to wish me well—and stayed.

Forward

Stowed in the closet of my office is a flight briefcase. It is packed with the navigation gear needed for a flight to anywhere in 1981:

      charts to everywhere in the world (now obsolete)
      flashlight and batteries (long ago corroded and useless)
      a variety of colored pencils
      my dividers for scaling distances
      two Jeppeson computers
      protractor for measuring angles
      scratch paper for computations
            (probably hotel stationery)
      air almanac for star positioning
      airline schedules
      plastic rain coat
      spare necktie
      roll of masking tape
      brochure listing London's best pubs
      navigation manual
      per diem expense forms

There are also the planned coordinates and routing that Varig gave us to fly from Rio de Janeiro to Algiers during the hijacking. Worldwide National Geographic maps and LORAN charts are next to two blue-printed charts for Los Angeles to Santiago that I had drawn by hand. An Inertial check list, a small sewing kit, and an extra pair of eyeglasses completes the list. A colorful "British Colonial Hong Kong" sticker on the end of the briefcase is silent testimony to the end of an era.

During my forty years of flying, I've watched aviation go from the China Clipper to the 747. I was in on the early days of air navigation when the centuries-old systems of seamen were being adapted and updated to travel by air. When I started, navigators sometimes had to stick their heads out of the plane to take star sightings. When I finished, complex computerized guidance systems had taken over. I've flown 15,000 hours, roughly the equivalent of fourteen round trips to the moon. I've slept in isolated shacks and luxury hotels and met fascinating people from Funafuti to Frankfurt.

I've watched the world grow smaller and smaller. When I was a child, my family crossed the country by auto, wondering at our amazing speed compared to the covered wagons of a few decades earlier. Today the same journey takes hours—or minutes if you're in a space vehicle.

Air travel has changed the world forever, for good or ill, and it could not have happened without the innovative, brave, and crazy aviation pioneers, the people who set out to fly somewhere—anywhere—with a star to guide them and a change of socks.

CHAPTER 1

Before Highways

> *Navigating from Tulsa to Los Angeles in 1926 before there were highways, road signs, motels, filling stations, or even Route 66*

My first navigating job was in 1926. I was seven years old.

We were living in Collinsville, a small town in northeastern Oklahoma, at the time, and my mother wanted to visit her mother in Los Angeles. My father, a merchant and local civic leader, decided it should be possible to drive to California in our hand-cranked 1924 Model T Ford. He interviewed several people in Tulsa who had already accomplished the trip.

The first problem to be solved was navigation. There were no highways. Each town had a series of radiating country trails going in all directions, primarily for the use of farmers bringing products to market. To go from town A to town B, you had to find the appropriate road. There were no numbered road signs to tell you where you were because all traffic was local.

To help us make our way, the Automobile Club of Southern California sent us a series of long cards. Each strip covered about a hundred miles, showing features like bridges, water tanks, railroad tracks, and anything else that would be helpful in "fixing our position."

Into the Unknown

We set out in May of 1926—my parents, my younger sister Jeanne, and me as navigator. The trip was almost canceled at the last minute because I broke my arm while doing acrobatics with a friend. But I was a very feisty seven-year-old, despite frequent illnesses, and I convinced my parents that all I needed was a sling to support the cast. Nothing was going to interfere with this adventure.

My job was to watch for visual landmarks from the back seat of our vibrating Ford. This was my first official experience as part of an Operating Crew. The Automobile Club strips had the recommended roads marked in yellow crayon, and I called out the next potential sighting while marking off our progress.

Preparations had been long and well-considered. My mother had decided skirts would be impractical for the rigors of cross-country driving, so she wore what today would be called a safari outfit. Indeed, we all looked as if we were on a safari. Our baggage included bags of water for the radiator and patching kits to repair the inevitable tire punctures. Filling stations were still rare, motels had not yet been invented, and eating facilities en route were sadly lacking. As a precaution against not finding a hotel at the end of the day, my father had rigged the front seat to hinge down into the rear seat to form a wide bed for him, my mother, and me. He contrived a cot for my sister which was supported by the four doors of the car.

Our first day's travel got us somewhere near Oklahoma City. The roads through the red, rolling hills followed the terrain up and down. Today's cut-and-fill level roadways were still years away. The route we followed was mostly unsurfaced single lanes, so it was a real treat when we encountered a few miles of gravel road.

As we rolled along, water-like mirages appeared on the road ahead of us. My mother had never seen one before and was alarmed. She knew that spring rains would mean fording small streams where bridges had not yet been needed—(horse-drawn vehicles didn't mind a little water)—and she begged my father, "Mark, don't drive into the water!" Our first night on the road, we slept under the stars in our makeshift beds.

I diligently marked off our progress: Amarillo, Tucumcari, Sante Fe, then Gallup. Between Gallup and Flagstaff, we stopped to see the Petrified Forest and the Painted Desert. They failed to live up to my childhood expectations. There were no standing trees in the forest, and the desert sure didn't look painted to me.

Prairie dogs would rear up on their hind legs in the middle of the single-lane dirt road ahead of us. We were on a collision course, but, just as it seemed we must hit them, they would scramble into their burrows a fraction of a second before we passed over them. Why they chose to have the entrances to their homes smack in the middle of the road I have no idea. Perhaps they figured they were there first and we were encroaching on their territory.

As we drove on, we noticed a series of signs marking the number of miles to Pie Town. All of us started salivating. With a name like that, there must be a good place to eat and gas to top off the tank. The rain was pouring down when we finally drove into town. It consisted of a single cabin-like structure. Fortunately, it sold food, but when my mother ordered milk for the children, the woman said, "We don't have any. If we had known you were coming, we could have gotten some for you." Automobile tourist travel was slight, and children were not the norm.

At Flagstaff, we encountered our first tourist cabins. For one dollar, you could stay in a small, individual wooden cabin with a wood-burning stove. It had a minimum of conveniences, but your car could be parked alongside, quite an innovation. From that point on, we were able to stay in tourist cabins every night.

At Williams, we diverted north to see the Grand Canyon. My left arm was still in a sling, which left me defenseless against "hostile" Indians, but fortunately we didn't encounter any. We did pass Navajo hogans where the women sometimes set up their looms beside the road. Travelers could stop to buy their rugs and blankets, or just to photograph them as they worked, surrounded by their playing children. A tip was expected in exchange for the photo, but I doubt that my father was overly generous.

Throughout New Mexico and Arizona, our route paralleled the Sante Fe Railroad. Instead of using dining cars, the trains had planned meal stops at the famed Harvey Houses. We took full advantage of these railroad restaurants.

Our Ford jounced its way through Williams, Kingman, Oatman, Needles, and San Bernardino. On the tenth day, we drove through desert land before reaching Pasadena. There I saw my first palms and orange trees.

I didn't realize it, but we had just traveled the future Route 66. It was officially established three months later.

Beyond Imagining

I'd like to say that I had a real sense of "arrival" in California, but I didn't. The entire journey, each day to and from Oklahoma, was more enthralling than any mere destination. Our first adventure, when we reached my grandmother's house in Los Angeles, was called by many "the scandal of the century."

As we drove up, the streets were full of newsboys hawking "extras." My grandmother had settled in the Echo Park area so she could be near the famous preacher, Aimee Semple McPherson, and her Angeles Temple Church of the Foursquare Gospel. Now, the glamorous faith-healer had vanished while visiting the seashore. Few people had radios then, so everyone rushed to buy the latest edition hawked by vendors in the neighborhoods. The whole city was eager for the latest speculation about what had happened to Sister Aimee.

Eventually, she walked out of the desert with a tall tale to cover what many thought was an extracurricular affair.

Los Angeles was the most sophisticated city I had ever seen. But even more overwhelming was the balmy climate. The weather was so mild and sweet, utterly unlike the harsh, violent summers I was used to back home. It elevated the senses and made everything seem bright and new.

Everyone at that time spoke of California as a golden land. It truly was. The skies were radiantly blue, the distant hills vividly detailed in the clear, smog-free air. Wonders glimpsed only in magazines and movies were suddenly before us. My mother's brother Alfred was a jovial Harvard man with many local connections. He introduced us to the revered Hollywood pioneer and cowboy star William S. Hart, boyhood hero of millions. Uncle Alfred also took us to Graumann's Egyptian Theatre, an architectural marvel, where we thrilled to dashing Doug Fairbanks in *The Black Pirate*. It was the first Technicolor film I'd seen, accompanied by a seventy-five-piece orchestra. Another adventure was an outdoor performance of Max Reinhart's epic passion play *The Miracle*, with a cast larger than the entire population of Collinsville, plus horses, camels, platoons of soldiers, and an enormous chorus. The show had been playing in Los Angeles for several years.

One day, we drove to Santa Monica along a country road called Santa Monica Boulevard. There, we had our first flat tire, and there I saw the Pacific Ocean for the first time. I stared at it and said, "Is that all the bigger it is?"

The Dressmaker and the Ice Skater

My maternal grandmother, Amanda Eugenia Nordstrom, was a small, spunky woman. She came from Sweden, traveling alone to America when she was fourteen or fifteen. Although she had a minimal formal education, she had learned dressmaking and tailoring in her native land. She was able to earn her keep in Nebraska by living with a series of farm families while sewing their clothes. Her only compensation was room and board. When the family was outfitted, she would move on to another farm and repeat the process. Still a teenager, she met and married Alfred Stridborg, a man in his forties and a photographer by trade. From that time on, this young girl dressed like a middle-aged woman, no longer pretty and young.

Homesteading was the thing to do at that time. After his marriage, Grandfather Stridborg gave up photography and tried farming on a homestead, but he wasn't very good at it. This failure must have been painful to him, for he had a colorful and romantic past. He had been a stone carver in the Old Country, a really good one. Somewhere in Sweden and possibly Russia, there are still cathedrals and public buildings adorned with stones carved by my grandfather. They have been in place well over 125 years—not a huge personal legacy, but more than most of us leave.

During military service, his name had been changed from Lindbergh to Stridborg for some obscure reason. In his youth, he is reported to have been quite an athlete. The family story is that, while performing on ice skates and doing calisthenics in imperial Russia, he was observed by members of the Royal Family. He was offered a job teaching physical fitness, but he declined, returned to Sweden, and emigrated to America. There he hoped to fulfill his ambitions to become a religious leader.

He was fortunate in his choice of a wife. Grandmother Stridborg managed to run their Nebraska homestead while she raised their five children. After my grandfather's death in 1921, she retired to Los Angeles. When I met her, advancing years had made little impression on her. She announced proudly that she had taken an airplane ride over Los Angeles the previous year. She remembered that the pilot's name was Art Goebel—a famous aviation pioneer of the time.

The Cast Comes Off

One evening, we set out, driving north over the Ridge Route to visit relatives in Turlock, a small Central Valley town 320 miles north of Los Angeles. By traveling through the night, we avoided the suffocating midday heat of the valley summer. From the Grapevine where the Ridge Route terminated, north to Bakersfield was an absolutely straight ribbon of road through the desert.

I had been born in Turlock just before World War I ended. In 1918, my father had gone into the army, and was stationed in Laredo, Texas, supposedly guarding our borders against raids from the Mexican bandit leader, Pancho Villa. My mother went to stay with relatives in Turlock to have her baby. My father got leave to come see us. While he was in California, the war ended.

During our stay in Turlock, my arm had mended sufficiently to remove the plaster cast. Within an hour, I tripped on a vine and broke it again.

By Night through the Desert

Our return to Oklahoma took us on the more southerly desert route, eastbound through Blythe, California. It had to be driven at night to avoid the extreme heat of the day. On the long, straight roads, my father put his foot down on the gas pedal, but there were few reference points in the vast open space to give any sense of speed. After one long stretch, another car stopped beside us at a filling station. "Do you know how fast you were going back there?" the driver asked. Model T Fords had no speedometer, so my father admitted he had no idea. "Mister," said the admiring driver, "there were times when you hit forty miles an hour!"

But this was the exception. We were making our way laboriously on a rutted sandy trail in New Mexico, when a Mexican man suddenly appeared ahead of us, waving his large hat. His signals indicated that a stop was mandatory. We were miles from any settlement, and my mother feared we were about to be "hijacked," the current term for a highway robbery. Father stopped the car. The man approached us, and suddenly, with a big smile, he extended his hand. It contained, not a gun, but several plums. He was offering us a gift, and gladly accepted a cigarette in exchange, which was exactly what he had hoped for.

The rest of our journey home was uneventful, except for a stop in Farmer, Texas to visit relatives, and two flat tires within several miles of Collinsville. Now seasoned travelers, we drove along, frequently comparing our great speed to that of the covered wagons that had crossed the Great Plains a few years earlier. I have often imagined those wagons drawn into a circle for the night encampment, with the lanterns going out one by one until the last one illuminated a domestic argument: "Not one more step! We're staying right here!" It's easy to understand how Denver got its start when you see the lofty Rockies looming to the west.

Several years later, we made this same trip in a 1928 Nash. Our driving time was cut to five days each way, with better conditions everywhere.

The Traffic Light

One day, my father heard that a traffic control device had been installed at an intersection in Tulsa. Everyone was talking about the red, green, and yellow lights that directed traffic without a police officer. Would drivers really stop at the red lights? Was it too much to expect people even to see the lights? What about the car behind? At each change of the light, there was a clanging bell that annoyed those who had to be nearby. The installation was only experimental, and could be removed if found to be impractical.

To experience this new scientific phenomena, we all piled into the family automobile and drove the twenty miles to Tulsa. It was a Sunday afternoon, and dozens of other curious motorists were there to see this new signal. Go on green, stop on red, get confused on yellow. My father was so fascinated that he went 'round and 'round the block in various directions so he could keep crossing that intersection. Then we drove the twenty miles back home, able to confirm from actual experience that the world was becoming automated.

The Gas Pump

In the 1920s, gasoline was dispensed at filling stations by hand operated pumps. The attendant rocked a three-foot lever back and forth to pump the fuel up into a glass cylinder. It was marked with lines like a thermometer, with zero at the top ranging down to about 10 gallons at the bottom. Then the attendant would put the nozzle in your tank, and gravity did the rest. You could stop at any point and were charged for the amount on the line level with the top of the remaining gas.

My father was a silent partner with a man in Poteau, Oklahoma who manufactured these glass cylinders. All it took was a lone Belgian workman who could blow molten glass bubbles into a cylindrical mold. The ends were cut off and sold as fish bowls, and the cylinders became gasoline dispensers. However, the product soon became obsolete. Production ended, along with my father's dream of becoming an automotive industrialist.

Oil!

We had returned from our trip to California via Tucson and El Paso before stopping in Young County, Texas at the town of Farmer. My father's uncle and family had settled the town many years before, thus the name. His descendants were living a marginal life on land that could barely support sheep. Still, my father had high hopes. He had studied geology briefly at Oklahoma University and felt there might be oil underneath the barren surface. His interest was more than scholarly. We could benefit directly because the inherited land was jointly owned by all the heirs, even those who didn't occupy the land.

A few years later, in the early 1930s, several oil wells were indeed brought into production, greatly improving the family fortunes. One of my cousins became a district attorney, and the others were substantial citizens in their community. The oil wells are still producing more than sixty years later.

Every Monday morning when we lived in Collinsville, a German widow named Mrs. Miller came to our home. She

Mark Farmer, age eleven. Even at this early age, it is obvious that aviation is a passion. In my right hand is a reproduction of the red World War I plane flown by Baron Manfred von Richthoften, the famed "Red Baron."

gathered up the week's soiled laundry and washed it by hand with a scrub board and tub. After the clothes had dried on the wash line, she'd iron and fold them neatly. For this day's work, she was paid one dollar, the standard wage even for store clerks. One Monday, after her work was completed, she approached my mother somewhat reluctantly. "I like you folks very much," she said, "and I appreciate the work, but, well, they've just struck oil on our little place, so I no longer need the money." Crude oil was selling for twenty five cents a barrel. These shallow wells probably pumped only $50 to $100 worth of oil a month, hardly great riches, but to poor people it was a measure of security they had never known.

There was so much oil being discovered at relatively shallow depths that some of us teenagers built small wooden drilling rigs, fashioning drill bits from auto parts and casings from drain pipes. We actually dug wells down to forty feet, making a big mess of our back yards, but never hitting any oil. The real payoff turned out to be the several dollars we were paid to clean up the yards.

Professional oil rigs were also simple, and an investment of several hundred dollars could get a share of the royalties if drilling was successful. Many of the workers got no wages. The drillers had little money, so instead they promised the workmen a monthly percentage of what was pumped if there was a "strike." When successful, the workers were often eager to sell their interests outright for a few hundred dollars quick cash. It seemed a much better deal than waiting for small monthly checks of $15 or so that could stop at any time if the well went dry.

My father's brother Clifford was a lawyer whose firm arranged leasing rights. Over the years, he bought up dozens of workers' shares for about $300 each. When my father died, Uncle Clifford arranged for my mother to make similar purchases. The monthly income from these small investments augmented our family finances greatly for over thirty years.

I Shoot Will Rogers

After Lindbergh flew the Atlantic alone in 1927, I could think of little else but airplanes. A dirigible flew over Collinsville, and barnstormers came through, giving sightseeing rides in World War I "Jennies." I was consumed with building scale-model airplanes in my garage workshop. To finance supplies (and an Iver-Johnson bicycle), I sold soda pop from a front yard stand by the main street.

In 1930, the Eastman Kodak Company celebrated its fiftieth anniversary by distributing gold-colored box cameras to twelve-year-olds. (Some of these cameras are now in museums.) The idea was to stimulate a new generation of eager photographers, and by extension, to increase future film sales. I qualified for a camera and proudly claimed it at the local ReXall Drug Store.

The next day, clutching my treasure, I rushed to nearby Claremore where they were dedicating an open-field "airport." It had been hastily thrown together, primarily because humorist Will Rogers would be there. He was a famous vaudeville, Broadway, and Hollywood star who had grown up a few miles away in Oologah. Rogers had persuaded his pal, aviator Wiley Post, and Post's navigator, Harold Gatty, to attend the dedication. Post and Gatty arrived in Post's single-engine Lockheed Vega, a plane they had flown around the world. Post, also an Oklahoman, was a dashing figure who wore a distinctive eye patch. He had lost the eye in an oil field accident, and used the insurance settlement to buy the Vega.

Will Rogers was an enthusiastic air traveler before there were any airlines. He wrote a syndicated daily newspaper column which was required reading for literally everyone in our area. Many people knew him personally, and enjoyed his grassroots philosophy and acidic prodding of politicians. It was a tragic day six years later, when we learned that Rogers and Post had been killed in a plane crash. Their pontoon-equipped Lockheed Sirius had gone down near Barrow, Alaska.

I still have two framed pictures of Will Rogers and the Vega which I shot that day with my gold box camera.

The Wright Stuff

My first musical instrument was a trumpet. Then my mother heard about a teenager who had a heart attack while playing one, and she made me stop. But I was eager to join the school band, so I took up the snare drum.

Our local high school had no budget for "frills." Basketball was played in a room that doubled as an auditorium, and football on a field without bleachers. But there was one bright spot: the school band.

It was the creation of the publisher of the local weekly newspaper, Claude Wright. As with all small town papers, the Collinsville News was the glue that made the community. Mr. Wright's journalistic style may have been wanting, but the way he ran his enterprise shows what just one person can accomplish.

Mr. Wright wasn't a philanthropist. Far from it. He had contrived a way to operate a school band while making a little money on the side. Each young participant was charged a dollar a month. The price included Mr. Wright's instruction and the use of his large and ancient collection of musical instruments and sheet music. When I started, only a french horn was left. As soon as a coronet became available, I switched.

Band practice was held an hour before school started, in an unheated room behind the gymnasium. Although Mr. Wright was not a patient man and his productions were pure corn, he counseled and led his ragtag group with great pride. The band did a lot for everyone who participated, talented or not. Fortunately, there were always several players who were very gifted. If such a fellow fell behind in his monthly dues, Mr. Wright kept him on for the sake of the group.

One of the highlights of the year was competing in regional competitions held on college campuses. Bands were classified by their town's population, so smaller communities didn't compete with larger ones who could afford more and

The Collinsville Band, proudly wearing the snazzy uniforms my mother devised. I am just left of the bass drum. Mr. Wright, in business suit, is to the right.

better instruments—exotic luxuries like kettle drums, oboes, and bassoons. In retrospect, I realize now that every band went home from these contests as a winner in some category.

Our band played for football and basketball games, and for any sizable gathering. When the circus came to town, we'd march in the circus parade and play outside the tent in exchange for free admission. And, by adding several violins, we became a church orchestra. I soon learned the violin and played at church services.

In 1931, everyone agreed it was time to get some uniforms. My mother volunteered. She made each boy a cape of cheap black felt with a red lining (the school colors). By adding twenty-five cent black berets, white duck pants, and black shoes, we became a uniformed marching unit, bursting with pride.

That same year, a large number of bands from northeastern Oklahoma were invited to Skelly Stadium in Tulsa where John Philip Sousa himself led them in some of his famous marches. Sousa stood atop a step-ladder as he conducted "The Stars And Stripes Forever," and I beat on a small snare drum. I have been known to say that I played in Sousa's band.

The extra $25 or $30 a month that Mr. Wright made from his band probably helped the newspaper survive. This arrangement was typical of small town practicality that would probably be impossible today with all the various regulations. Still, the school got a band, the taxpayers saved money, no one accepted any charity, and teen crime was nonexistent.

Recent studies have shown that young people who get musical instruction do better in their academic studies. This was certainly true in Collinsville. Many of the members of the school band went on to college, though this was partly because there was no work anyway. Prosperity was still hiding "around the corner."

Mr. Wright's practicality extended to his own four sons. He made sure they were able to operate an old fashioned hot-lead Linotype and sheet-fed press. All of them earned their way through college as Linotype operators.

Six (or More) Marks

The Farmer family had been dry goods merchants for several generations. Around the turn of the century, they left Joplin, Missouri to resettle in Indian Territory. Everyone in the family was small in stature, which is probably why they decided to be merchants rather than fighters or farmers.

My father was named Mark Kearns Farmer. So was his father, grandfather, great grandfather, and who knows how many others before them. (I am also Mark Kearns Farmer and so is my son.) Small but wiry, father tried to make up for his size by being an overachieving athlete. He enjoyed success in track and baseball, pitching both right- and left-handed. He and his brother Clifford were excellent marksmen and avid fly casting fishermen. One time my father cast a plug with several fish hooks and caught two fish at the same time.

I'm not certain when my grandfather died, but while my father was attending Oklahoma University, he was called back to Collinsville to take over the family store. It became Farmer Bros.—plural—though my Uncle Opal left the management to my father. Their two other brothers were never involved.

As the new operator of Farmer Bros., my father made frequent buying trips to Kansas City. In those days, the traveling salesmen, known as drummers, provided their customers with news of which stores were succeeding and which weren't and therefore might be for sale. Failing businesses were also a good source of potential employees. Through these drummers, my father heard about Ruth Stridborg, a young Swedish Nebraskan, who had been apprenticed as a milliner (ladies' hat maker) in Kearney. He hired her, and, several years later, they were married.

The Good and Bad of Small Towns

Nobody locked their doors in Collinsville. If you were going to be gone a number of days, you might lock up to deter larcenous outsiders, but that was scant protection. Nearly every door in town could be opened with the same skeleton key. Mostly, people trusted each other.

Not that lawlessness was unknown. Bank robberies had become so popular that the newly established bank erected a protective iron gate at its front door. You had to be recognized and admitted by the manager. And there were occasional disagreements, sometimes fueled by home-made Prohibition liquor. By day, most merchants kept loaded shotguns in their office areas so they could take care of any emergency themselves.

From 1920 to 1933, all alcoholic beverages were illegal in the United States. The "Great Experiment" or Prohibition as it was called, turned many otherwise honest citizens into lawbreakers. Often, when groups went into the wooded areas for picnics, "spotters" would appear a short distance away. They were just making certain that no one got too close to their unlawful enterprises.

Once, the empty house across the street from us was rented, but no one ever saw the occupants. There did seem to be visitors who stopped their cars in front for short periods in the evenings. After several weeks, the house was raided by lawmen who found a whiskey still. They smashed it with sledgehammers and dumped the liquor and grain mash on the front lawn. It killed the grass and left a telltale odor for several days, a warning to all.

The only local lawman in Collinsville was Mr. Singleton. He was all the town needed for protection. At night, he patrolled the downtown on foot, walking the alleyways where farm families parked their horse-drawn wagons and surreys while they purchased supplies and got current with local gossip. Stores stayed open until ten o'clock on Saturday night. The kids could see a ten-cent cowboy movie while their parents shopped and visited.

Officer Singleton did not wear a uniform. He was a real frontier, no-nonsense authority with a large black hat, bushy mustache, and the slow walk of a large man in his fifties. He was known to be an excellent marksman.

My father never kept any money in the iron safe in his store, but one Sunday morning before daylight, two men broke in. Unable to open the safe, they decided to cart it away with them. Of course, Singleton was right there to make an arrest. The thieves fled. One was shot, and the other escaped on foot. Singleton ran two blocks to his home, jumped in his car, and went looking for the fugitive. On the edge of town, the burglar tried to hitch a ride with a passing car. It was Singleton, who obliged. The next day, my father took up a collection from the other merchants to reward Singleton for his vigilance. The sixty dollars they presented to him was probably a month's salary.

However, there was a darker side to this small-town atmosphere of fellowship and trust—a certain morbid mentality. For example, the frontier practice of the wedding night shivaree was still popular. Rowdy townsfolk would converge on wherever the couple were staying, making as much noise as they could. Then they'd drag the newly married man from his bed and chain him to a street lamp or tow him in a tub behind an automobile before throwing him in a nearby river. This humiliation was considered "good fun."

There was also an ad hoc Ku Klux Klan. It started in 1921, after a big fire in nearby Tulsa left many Black families homeless. A rumor spread that there could be an "invasion" of refugees into Collinsville. A local group quickly formed and got the word out that, "the sun will never set on a N——- in this town." And it never did.

Untold Wealth

The Taylor family lived next door. Mr. Taylor worked for the Santa Fe Railroad as a small-bridge construction foreman. He probably earned $60 a month, and was seldom able to come home. There were two Taylor boys my age, plus some grown children and their offspring. The whole family lived in a four-room, rented frame house which was heated by a wood stove and had no inside plumbing.

The two boys augmented the family income by delivering magazines to subscribers. Mrs. Taylor was never seen shopping down town, a block away. I don't think she had presentable clothing to go out. But she was constantly reading to her children, a luxury I did not fully appreciate at the time.

Another bounty the family enjoyed was qualifying for free passes on the Santa Fe Railroad. A few years after we had made our automobile trip to California, Mrs. Taylor took her brood to California too. Not having money for the Harvey House restaurants, she packed enough food for the entire round trip! I imagine those kids had been thoroughly briefed on what to look for all along the way. To me the Taylors represented unforgettable, old-fashioned character and determination.

The Imaginary Invalid

My mother was overprotective of my sister and me. When we "got sick," she would call Dr. Wilkes who made house calls for one dollar. He always left a small envelope of pills, and I was allowed to choose the color. Eventually I had sampled his entire collection. Occasionally, my mother was instructed to dose us with a teaspoonful of coal oil (lamp oil) mixed with sugar, or a spoonful of whiskey in a glass of water, or Vicks salve mixed with castor oil. Because of Prohibition, the whiskey had to be obtained by prescription from the local ReXall Drug Store.

Once I was kept in bed for months so my "heart would not be strained." There was never a name for what was supposed to be wrong with me. No tests were made. "Avoid his turning over in bed, and maybe he will survive," Dr. Wilkes told my terrified mother. Mother was told to boil beef and squeeze the juices out with a colander. The resulting liquid would be "more digestible than bulky meat fibers." I missed an entire year of school. In the spring, I was allowed to get out of bed, but was so unsteady that it was several days before I could walk with any confidence.

It was during these periods in bed that the National Geographic Magazine and stamp collecting became my companions. There was never anything wrong with my basic health, but looking back, I realize that I grew up in sealed rooms with unvented gas heaters. No wonder strange things happened. The high level of fumes must have been far worse than the "cold, harsh winter air" which was not allowed to enter.

Rise and Fall of a Midwestern Town

My paternal grandparents had decided to settle in Collinsville rather than Tulsa, an adjacent town, possibly because the Guggenheim zinc smelter on the edge of town was a big employer. The smelter was located near strip-coal fuel, and the zinc ore was brought by rail from Missouri. During World War I, at full operation, the smelter made for a prosperous town.

After the war was over, the smelter ceased operating. Unemployed workers left, and a steady depression became a way of life throughout the 1920s.

The town's rough streets were paved with red bricks which made for miserable childhood skating and bicycling. The bricks had been produced by a local brick kiln, fired with coal from the nearby pits. Retorts for the smelter were made by the same facility, but it was forced to close down when concrete roads began to replace brick ones.

Collinsville had four mercantile or "dry goods" stores that eventually served about 1200 residents and nearby farming families. When the smelter failed, two of these stores closed. My father's store, Farmer Bros., stayed open.

Then the stock market crashed in 1929. My father was fortunate not to own any stocks, but Uncle Alfred in Los Angeles lost his brokerage job. Grandmother Stridborg had invested with him and lost her retirement savings. She should never have been on margin, not understanding the risks, but everyone was doing it and her son Alfred, after all, had a Master's degree in Business Administration from Harvard. Surely he couldn't give her bad business advice.

My father had kept his working funds "safely" in the local bank. But, several months later, Mr. Coburn, the local bank president, walked into our driveway early one morning and, in tears, rapped on the window screen of my parents' bedroom. He said, "Mark, we aren't opening the bank this morning. We are broke." I don't know how much money was lost, but several merchants immediately went out of business. From then on, Mr. Coburn seldom left his home, as though he were hiding from the whole community. I took it hard. My life savings of $18 were in that bank, a heavy loss for an eleven-year-old. Fortunately, my father had excellent credit with the wholesalers, and he was able to keep going.

With the failure of the last bank in town, local business nearly came to a standstill. Only the nearby Osage Indians had any money to spend, the income from their oil rights. A clothing sale of $100 to an Indian family saved many a business day at my father's store.

Unemployed workers could move on to look for work elsewhere, but anyone who owned a business or a home was stuck. It was nearly impossible to sell them and start over somewhere else. In this, Collinsville was not unique.

The Ticking Time Bomb

Because an economic time bomb was ticking, it was considered akin to treason to spend one's money outside the community. The local newspaper would report with implied disapproval that "Mrs. So-and-So motored to Tulsa for shopping." Then as now, there was great concern that the growth of chain stores would put local merchants out of business. A fellow named Henderson in Shreveport, Louisiana used the radio to warn of the chain store menace. Small merchants all over the South sent in their contributions to keep his program on the air. "Send me a dollar," he'd cry, "and I'll mail you a pound of Hello World Coffee. We'll continue to fight the chain store menace, and dang their buttons!" Now, a dollar for a pound of questionable coffee was an enormous price, but it was for a "good cause." Just to be able to listen to a radio station as far away as Louisiana was an event in itself.

By early 1933, the Depression was still getting worse, even though $60,000 of federal money had been appropriated by the Works Project Administration (WPA) for hiring local people to improve the school building and some other public works. An unused brick school building on Smelter Hill was dismantled by hand, and the bricks were moved to the playground of Central School which was still in operation. A Mr. Goldsmith, whose mercantile business had failed, was employed by the WPA to clean these bricks so they could be used for some new grade school buildings. This man in his fifties who had never done hard physical labor was leading the WPA group by example, and he couldn't even afford work clothes. He worked each day in a suit.

The Funerals

One morning when I was in the eighth grade, the school superintendent came to take me home. Nothing was said, but I knew something very unusual must have happened. When we reached our house, my mother told me that my father had died while opening his daily mail, apparently of a heart attack. He was thirty-nine. The year was 1931. The whole town closed down for his services.

During the five days we waited for Grandmother Stridborg to arrive from Los Angeles by train, my father's body and coffin were kept in our front bedroom so friends could visit and pay their respects. My sister Jeanne remembers that Mrs. Oberhaus told her that "Jesus needed your father more than we did." I heard Mrs. Blevans say, "It's too bad that the husband and father had been taken, and not one of the children." I had never thought of it as a negotiable proposition.

Grandmother Nora, my father's mother, lived a block away from us, midway between his store and our home. For more than a decade, we had passed her house daily, knew what she looked like, but had never spoken to her. Each Saturday evening, she would arrive at the store and sit in a chair near the pattern counter, where women gathered to thumb through the latest pattern books. There she held court in the store that once had been hers, a tiny, sharp-featured figure dressed all in black.

Her relations with my father were amiable enough. He even sent her to Colorado each summer so she could escape the oppressive Oklahoma heat. But when my mother, sister, or I were in the store or passed her on the street, she never acknowledged our existence. It was an odd situation, never discussed, but we accepted it as children accept many of the adult mysteries around them. Later, we learned that my father had rejected his mother's choice for his wife and married "beneath him." The first and only time we ever sat with my grandmother was at my father's funeral.

The day my father was buried, the entire town shut down. The 300-seat church was filled to capacity, and there were several hundred people outside. He hadn't held political office, but was thought of as a "king maker" and a power "behind the throne." Candidates for office had regularly sought his support, which, when granted, virtually guaranteed their election. Our congressman sent my mother a letter of condolence.

Soon after the funeral, Grandmother Nora became bedridden. My mother spent countless hours tending to her needs before she died, never offering a word of thanks.

If You Can't Give It Away, Sell It

My mother probably had an eighth grade education at most. She knew little about bookkeeping or the mechanics of business, but she possessed a remarkable innate knowledge of how to buy and sell merchandise. After my father died, out of necessity, she became the backbone of the family business.

One of her first brainstorms was how to deal with the accumulated junk in the warehouse rooms above the store. For years, obsolete and unmarketable merchandise, including hundreds of pairs of ladies' high button shoes, had been relegated there. My mother reasoned that, at the right price, farm women might be able to use them. She rented the vacant First National Bank building, hired a Mr. Burchfield as manager at a dollar a day, and opened the Collinsville Bargain Center. It was stocked with all the uninventoried leftovers, which she supplemented with cheap merchandise obtained in Kansas City. The high button shoes were priced at ten cents a pair, and they sold like the proverbial hot cakes.

Our lone competitor in town also had an inventory of unsalable high button shoes. Seeking to outdo us, he offered them free under a big sign that said, "Save Your Dime." His strategy backfired. Folks thought he was implying they couldn't afford ten cents for a pair of shoes, so they patronized the new Bargain Center instead. Because this was supposed to be a new business in town, we had to smuggle in the old inventory from Farmer Bros., sneaking it down an alley at night.

The Mysterious Uncle

My father had three brothers. Uncle Clifford was a prominent lawyer in Tulsa. Uncle Opal worked as a clerk in the family store and lived with Grandmother Farmer. The third brother was somewhat mysterious. Throughout my early years, Uncle Paul would appear periodically. He was a pleasant fellow in his mid forties who always wore a suit and tie. He had little to say, was unmarried (as far as we knew), and had no apparent job affiliation. When we had visited Los Angeles, he showed up for a family group picture, then disappeared. The only thing I ever saw him do was play checkers at the firehouse. My mother said once that he played cards in the Chicago area. No one was certain what his activities were, and no one probed.

After my father's death, mother took over management of the depression-ridden store. Uncle Opal left to became a liquidator of bankrupt mercantile stores, buying them for ten cents on the dollar and then selling the inventories at cut-rate prices throughout small towns in Oklahoma. There seemed no end of bankrupt stores, but nevertheless it was a marginal business. Uncle Clifford probably took over Grandmother Farmer's support for the few months she outlived my father.

The third funeral that year was for Uncle Paul. He committed suicide with my father's shotgun. Only family attended the services. Years later, I realized that Uncle Paul had probably come to Collinsville to borrow money. After my father's unexpected death, Uncle Paul was too embarrassed to ask my mother for money. Or perhaps he did, and she turned him down.

God and the Depression

When Collinsville was in it's heyday, there were six churches to serve the community. But, as the population dwindled, only the Baptist church could afford a full time minister. This was happening in many smaller American communities, and a number of them decided to consolidate the various local congregations into a single multi-denominational "Community Church."

Finally, the nearby Christian Church fell seriously behind in mortgage payments on its nice, tall brick building. It was obvious that something had to be done. A minister was brought in from Minnesota, and the Christian Church became the Community Church, a flagship building for local churches. The Presbyterian church joined the group, and its former property was turned into a recreation center for the town. I remember their two dirt tennis courts, ringed with a fence made from used oil-well pipe. Not all Methodists broke away, and no Baptists joined this new community effort, but there were Catholics and several Jewish members. The minister would baptize by immersion or sprinkling, whatever one desired. Those participating had a real sense of getting the community out of its depression-ridden doldrums. It was a great experiment that seemed to work, even without money. Finally something positive was happening.

Then, on a winter Sunday morning early in 1933, a defective coal stove started a small fire in the newly refurbished Christian Church building. The volunteer fire department stood by, helpless, because the water mains were frozen. The fire quickly consumed the building and the dreams of a struggling community. The movement regrouped and continued to meet at a local movie house.

A few months later, we left for California.

Starting Over

Just when it had seemed impossible for things to get worse, Oklahoma had been hit by the "dust bowl" phenomenon. Years of uncontrolled cultivation and cattle-grazing in the plains states had made the topsoil vulnerable to the prevailing winds. When drought hit several years in a row in the early 1930s, the relentless wind became a terrible enemy, carrying off soil in thick, choking clouds. No crops would grow, and farm after farm was abandoned. The economy of the towns was further devastated, and Farmer Bros. faced the possibility of closing.

We knew from newspapers and letters from our Los Angeles and Turlock relatives that the Depression had not been as severe in California. With the irrigation systems there, it was unlikely to become a dust bowl. My mother decided that it was time for the family to move west.

She sold Farmer Bros. for $1,500. The new owner got a $23,000 inventory and a two-story commercial building for this price, but mother was lucky to get it. We had disposed of everything we couldn't pack in the trunk of our car, getting just a few hundred dollars for all our furniture and household goods. Then, just as we were ready to start for southern California, we got the news. A disastrous earthquake had hit the Long Beach area. My mother was now very apprehensive about resettling in an earthquake area, but it was too late to change our plans.

So, on May 30, 1933, we climbed into a brand new red Chevrolet and started west. A young man named Benny came along with us to help with the driving. It was his method of getting out of Oklahoma. Tucked tightly in my mother's purse was $5,000 cash from an insurance policy. Not all Oklahoma migrants were as destitute as the "Okie" farm workers chronicled so poignantly in John Steinbeck's The Grapes Of Wrath.

This time, we reached Los Angeles in just two and a half days, one-fourth the time our 1926 trip had taken.

We rented a small apartment for $20 a month on a modest cul-de-sac near the Echo Park area where my grandmother lived. From there, mother scouted the Los Angeles area for a business opportunity. She was offered work as a saleslady, but at wages that would never have supported a family.

After a fruitless search, we headed north to Turlock where her sister lived. This made a lot more sense. Turlock was an agricultural town of less than 5,000 people. It had excellent schools and no real Depression scars.

My mother's store in Turlock, California, late 1950s.

Mother soon found an unoccupied space on the main street that had been a ladies' ready-to-wear store. Fortunately, it already had covered dress racks and fitting rooms. There were even enough fixtures and display cases to start a modest ladies' store, providing we did the refurbishing ourselves. The rent was $50 a month.

For the next month, my mother, sister, and I scrubbed, painted, and redecorated the space in a modernistic art-deco style. We added some smart, new chromium-pipe chairs to complete the effect. Then mother drove 100 miles north to San Francisco, where she bought $1,500 worth of merchandise. She hired several young women at twenty-five cents an hour, and it was time for the Grand Opening. The "Ruth H. Farmer Ladies' Ready to Wear" was ready for business. Mother let it be known around town that she had come from "back east" which hinted at the smart fashion salons of New York, Boston, and mainline Philadelphia. Her first promotional event was a free fashion show for which she did the commentary.

In the next few weeks, sales ran $30 to $40 a day, with a good Saturday bringing in $100. This was marginal, but, combined with the small monthly royalty checks from the oil wells, she was able to keep body and soul together as she built up the business. Ten years later, she was grossing between $100,000 and $150,000 a year.

When World War II started in 1942, a small defense plant was set up in Turlock. It generated the first disposable income that most of the workers, many of them women, had ever known. Automobiles and hard goods were no longer available—manufacturing was limited to military needs—so these women used this new money to buy nicer apparel. My mother made sure she got her share of the business.

For the next three years, she prospered, saved, invested, and improved both her store and home. In the 1950s, the bank building burned down, taking her store with it. With the insurance money, she reestablished in a different location and operated successfully until her retirement in 1960. She had invested wisely, and was able to enjoy a comfortable retirement.

George

Throughout her years in business, my mother used the janitorial services of George Yardy. He was a working-class Englishman, a short, stubby-bearded man who made his living washing storefront windows and polishing floors with a tricky-to-operate machine. Mother appreciated his industriousness and unfailing reliability. He had a rolling walk as if one leg were shorter, and one of his eyes was distended and uncoordinated with the other.

George carried his equipment around in an old Chevrolet panel truck, which he also used to pick up anything offered him that "might come in handy some day." His home was on the edge of town where he could keep goats and store the scrap metal and castoffs given to him. Often his young boys would help with his cleaning jobs.

I never saw George wearing anything but worn bibbed overalls and a cap. One day, he wandered into a local automobile dealership and stood there, admiring the shiny new cars. The salesman ignored him as long as he could, then decided to have some fun with him. "Mr. Yardy," he said jovially, "how would you like to buy this new panel truck?"

"Yes, I'll take it," said George.

"And how would you like to pay for it, Mr. Yardy?" asked the salesman, going along with the joke.

"Cash," George replied. He fished a wad of money from a pocket in his overalls and counted out the precise amount. Then he moved on to another car. "And I think my wife would like that one over there." Another sale, another overalls pocket to be exploited, and a richer and wiser salesman who didn't even have to take a trade-in. When World War II started, George's scrap metal had real value. His son went to medical school and became a doctor.

Size Didn't Count

I started my second year of high school in Turlock. Mother was in business, and our situation, while tenuous, was not hopeless. I found that I took after my father, who, though small, had been very competitive in baseball and track events. Contact sports didn't appeal to me—I weighed just 125 pounds—so I concentrated on tennis. I always had a first class racquet with the best strings.

Mr. Livingston, the chemistry and physics teacher, liked to play tennis after school, so he became the unofficial school tennis coach. He also became a sort of adopted father to me. I found that I was pretty good at chemistry, physics, and tennis. A few students were interested in tennis, but I got so good that no one dared to challenge me. I was the champ, which was a great feeling for a pint-sized kid.

My only peer competition came when we played matches with surrounding towns. My first adversary was a large, good-looking fellow from the town of Patterson. He had more skill and experience, but I had fierce determination. I won. By the following year, I was at the top of the ladder, and was never defeated in high school competition. The myth of my childhood heart problem was completely forgotten.

The Cannery

The nearby Hume Cannery employed many high school students for the apricot- and peach-packing season. I was among them. Most of the workers were local folks, many from nearby farms who were supplementing their incomes. Some had been returning to the same assignments for over forty years.

When the season started, everyone would go and stand outside the cannery early in the morning. The foreman would come out and point to the people he wanted. Those with former experience were chosen first. New people might wait many days. Eventually I got picked.

In 1934, the pay was thirty-one cents an hour, or you could do piecework that sometimes brought fifty to sixty cents an hour. The work was grueling and hot in the valley summers, but I managed to put away several hundred dollars for college. Over five summers, I accumulated over a thousand dollars.

The $3,000 Sandwich

Between 1933 and 1942, mother made a meager living, but she was constantly upgrading her store and increasing her inventory. At first, the store hadn't carried any hosiery, undergarments, or accessories. She hadn't wanted to gamble too much all at once. But slowly it expanded. In 1935, Ruth H. Farmer Ladies Ready to Wear moved down the street into quarters next to the Bank of America whose air conditioning exhaust was diverted into the store for cooling.

At times, customers would request items like fur coats which she could not afford to stock in quantity. Mother kept a list of ladies who had not quite found what they wanted in her store. On her next month's trip to the markets in San Francisco, she would choose coats and items specifically for these customers. When she got back, she'd phone them and say, "I found something in San Francisco that looked just like you. I've got it here for you now." Sometimes, she'd call the husbands who were happy to surprise their wives with their good taste.

On these buying trips, she always wore a lady's business suit, white gloves, and a tasteful hat. (No lady left the house without hat and gloves in those days, and "get your hat" was a common synonym for "let's go.") My mother would carry the merchandise back with her in the trunk of her car to save shipping charges. On one trip, she stopped for dinner in Tracy. While she ate, everything was stolen, a devastating financial blow. She called it the most expensive chicken sandwich she ever ate.

Gossip

In the 1920s and 1930s, there was not much in the way of sex education, especially in small towns. Any discussion of alternate sexual preferences was usually done in excited whispers. One evening, my mother was attending a social event with a group of her lady friends. The gossip eventually got around to a certain lady in the community who was suspected of being a lesbian. Several in the group murmured observations that seemed to support this theory.

My mother held her tongue while her neighbor was more and more maligned. Finally, she stopped the discussion. In a firm, authoritative voice, she told the gathering: "I know the family well. She's not a lesbian. She's a German."

All Wet

The Turlock Irrigation District celebrated it's fiftieth anniversary in 1937. This news was hardly of any significance to the nation as a whole, but to me it represented the best of the "American spirit." The Central Valley of California had been flat and waterless, good only for dry-land grain farming. It took thousands of acres of land to make a profitable enterprise. The large dry-land ranchers realized that if they irrigated the land, a small plot could provide a good living, and they could become quite wealthy by selling off these small parcels.

The ranchers formed a group, bonded themselves for several million dollars, and built a dam with water-powered electrical generators in the foothills of the Sierras. Irrigation canals and power lines brought water and electricity thirty miles to where it could be distributed to users. They showed what a few people can achieve if they work together.

With water, the valley flourished. Armenian immigrants came to grow their melons, and Japanese to do intensive vegetable farming. Portuguese from the Azores excelled at dairy farming, and Swedes from Minnesota and Nebraska came to raise chickens and fruit. Land was inexpensive, water and power were cheap, and the dry-land farmers became wealthy. Railroads were built to haul the produce to markets. Small towns grew up in the newly prosperous areas. Canneries packed the fruit, and melons went to the East Coast via iced refrigerator cars.

Other areas observed the success and wanted to duplicate it. But somewhere along the line, the concept of what the original farmers had accomplished was eclipsed by the notion that the government should build the dams and power plants. Larger dams were being built, the dirt canals cemented, and power remained inexpensive. The pioneer spirit to achieve was lost.

First Across

One day in November, 1937, my mother let me drive her new red Chevrolet to San Francisco for a very special event. Until that day, the only way to get from Oakland to San Francisco was by ferry. But this was about to change with the opening of the new Bay Bridge. We had been following its construction for four years, watching from the ferry boats whenever I came with my mother on one of her buying trips.

I invited several friends to share this memorable historic event. One of them was James Mitchell, a neighbor who later became an actor and dancer. (He was on Broadway in *Brigadoon*, *Paint Your Wagon*, and *Carousel*, and danced the role of Curly in the film version of *Oklahoma!*

We left Turlock very early in the morning, hoping to be one of the first in line to cross the new bridge. As we approached the toll gate, we saw that there were about forty cars ahead of us. We parked in the line and walked to the toll collection station where a canopy had been set up for the usual dedication. Ex-president Hoover was there, and the Governor and several mayors. After the speeches, the toll collectors marched in their new uniforms to their assigned stations. They looked so smart, I wondered if we should salute them as we paid our toll.

The real thrill was to be in the first group of cars to set out across this newest engineering wonder. Photographers were high on the eastern bridge tower to capture the first surge of cars. They claimed they felt the tower move slightly as it reacted to the first traffic flow.

As soon as we got to San Francisco, we turned around and drove back over the bridge. (In those days, cars traveled in both directions on the top level, while streamlined commuter trains ran on the lower deck.) Then we all drove home, tired and happy.

Another bridge was still being built on the other side of the city, crossing the narrow "Golden Gate" entrance to the vast San Francisco Bay. There was not much in the way of population at the other end of the bridge, but it was assumed that easy access would encourage growth in Marin County to the north, and that the bridge might ultimately be used at its capacity. (Little did they know!) The following May, the Golden Gate Bridge opened and became the symbol of San Francisco, greeting the flying-boat Clippers as they arrived at Treasure Island, beside the Bay Bridge we had just crossed.

War Clouds Gather

It was always assumed that my sister and I would go to college. My mother's brother, Uncle Alfred, had graduated from "Cal" and gotten his MBA from Harvard. I chose the University of California at Berkeley. My sister, two years later, went to UCLA, but eventually graduated from Berkeley.

At the beginning of my junior year, the war in Europe was looking ominous. I was studying mining engineering, waiting tables in a girls' boarding house, and sharing a small apartment with four others. Tuition was $27 a semester, and my share of the rent was $12 a month. By the end of the school year, I was sure that the U.S. would soon be involved in the European war. The army would soon be drafting men, and I assumed I'd be called up before I could finish my senior year. All I could think of was the carnage in the trenches of the last war, and I didn't want to be a "walking soldier." I decided that the best way to stay out of the trenches was to make myself invaluable in military construction.

In Europe, Hitler had already annexed the Sudetenland, invaded Poland, and begun the Battle of Britain. War in Europe had been raging for more than a year. The U.S. government was boosting aircraft production to supply the British with desperately needed military airplanes.

I loved drafting and descriptive geometry—that is, finding graphical solutions for problems. I decided to try my luck in the burgeoning, "high-paying" aircraft industries in southern California. If war came, I could be useful. If we got laid off, I'd have saved up a few thousand dollars to go back to school.

In June of 1940, I took a train south to Los Angeles. I had $15 in my pocket to tide me over until I got a job. Uncle Alfred met me at Union Station and offered his home in North Hollywood as temporary billeting. It took me only a few days to land a job at Lockheed in Burbank.

CHAPTER 2

Wings

> *Lockheed tools up to make Hudson Bombers and P-38s for Britain as war clouds loom in Europe. Then the Japanese bomb Pearl Harbor.*

The first day I walked onto the floor of Lockheed's Burbank plant, the noise nearly knocked me over. I was standing in the largest indoor space I'd ever seen, the size of several city blocks. Dozens of airplanes were coming together, swarmed over by thousands of workers.

I hoped to start as a draftsman, but they put me to work in the inspection department. The pay was fifty-one cents an hour. I was assigned to a shop in the heart of a huge complex where Hudson Bombers were now being made for the Britain's Royal Air Force. Coincidentally, it was the same shop where Wylie Post's Vega and Sirius planes had been built.

My first job was to make sure that sufficient zinc chromate had been sprayed onto aluminum-sheet metal parts. One day, another Inspector, an older man, brought me a problem. He was holding a blueprint for some thin metal strips that were bent into a circle. Only the diameter of the circle was indicated. "How long should the strips be?" he asked me. At last,

I had an opportunity to put my years of engineering training to use! Actually, I had learned the formula back in seventh grade: $\pi d = c$ or pi times diameter equals circumference. And this senior guy didn't know it. My hopes of promotion took a big jump upward!

Most of the people I was working with looked on Lockheed as just a job. They were excited about finally making some money after years of poverty during the Depression. I liked the money too, but I was also fairly sure the U.S. would soon be involved in the European war.

A few months later, several of us were transferred to the Loft and Template Department where our knowledge of drafting was put to better use. It was a heady time. Even though we weren't officially at war yet, I was already part of the "war effort," making the world safe for democracy.

Lockheed was expanding at a fast clip, so overtime was available and advancements were rapid. I bought a 1936 Ford with my new wealth and found a room to rent in nearby Westwood Village.

The Hudson Bombers

The Hudson bombers were actually an adaptation of a Model 14 civilian transport plane. Courtland Gross, brother of Lockheed's Chairman of the Board Robert Gross, had been in England and learned the British desperately needed a patrol bomber. He mentioned it to his brother who leaped into action. Lockheed took a Model 14 transport plane, put a gun turret on top, camouflaged it, and renamed it the "Hudson Bomber." They had been tooled up to produce the Model 14, but not in any volume. Now the factory was expanded and went into mass production. They even bought the private Burbank Airport so they'd have a place to test-fly the finished bombers.

Each day's newspapers proclaimed the need for skilled people. Lockheed, Douglas, Martin, Boeing and many other small aircraft companies were suddenly bursting out of their small quarters—in some cases, just a couple of hangars. With

The P-38

the impending war, all these companies grew at a tremendous rate. Eventually, Lockheed had 50,000 workers. After Pearl Harbor, the facility was camouflaged and anti-aircraft batteries were put around it, but all this happened after I had left.

The P-38s

One day, our boss told a half-dozen of us to report to the 3-G Distillery, a bonded-whiskey warehouse several miles from the main Burbank plant. We were to say nothing to our fellow workers, so we knew something was up.

When we got to the Distillery, we found a miniature tooling factory for the innovative new P-38. We would be making the tools—the dies, jigs, form blocks, drills and drill jigs—that would eventually be used to produce the plane parts in quantity. Our job was to convert a blueprint into a pattern or template that could be used to manufacture these tools for the first P-38 production aircraft. I felt privileged to be part of this project.

Soon after, I got another big morale boost. A bunch of us were chosen to go for supervisor's training at the Walt Disney Studios. This isn't as odd as it sounds. They had just finished Fantasia and had some unused rooms available a few miles away. The Disney organization wasn't well off at the time, and agreed to rent the space to Lockheed. Our training there made us feel recognized.

The first experimental prototype of the P-38 had recently crashed in a cross-country flight. Here, away from the hubbub of the main facility, the revised P-38 was being carefully retooled in preparation for mass production. It was a relatively small facility with several hundred employees.

At the Distillery, I worked the swing shift from 3:30 p.m. to 12:30 a.m. It paid fifty-six cents an hour, five cents more than I had been making, and it allowed me to take classes at UCLA in the mornings. After a few months, we were transferred back to the main plant where I supervised a group of twenty. I got extra money for "coordinating" with the day shift. This meant that I'd come in an hour before the day shift was through, find out what they wanted done, and act as liaison between them and the swing shift. Working ten hours a day, seven days a week, I was able to make over $100 a week. Even before Pearl Harbor, toolmakers were already expected to work seven days a week. You can't make planes without tools.

Even sixty years later, in this day of space flight, the P-38 remains awesome. With its high speed and altitude capabilities, it was one of the most distinctive, technologically advanced, and beloved aircraft of the war. The Germans called it the "fork-tailed devil." The Japanese said it was "two planes but just one pilot." It was a day bomber, a night bomber, a torpedo bomber, a photo-reconnaissance plane, and an air ambulance. It was first used over New Guinea, stopping the Japanese invasion. It went on to outstrip German Messerschmitt-109s over Europe and North Africa.

Connies and the Millionaire in Tennis Shoes

Howard Hughes had controlling interest in Trans-Western Airlines. (When it became an international airline, the name was changed to Trans World Airline, TWA.) At that time, all commercial aircraft was at the mercy of local weather. Hughes wanted a pressurized passenger plane that could fly at high altitudes above the weather, nonstop from coast to coast. The Constellation or "Connie" was developed in the experimental department at Lockheed. When I heard that the millionaire himself would be visiting, I expected someone in elegant clothes, maybe even gloves and a cane. However, Mr. Hughes was invariably dressed in sloppy clothes and tennis shoes as he crawled and clambered over the emerging model.

The Connie was still experimental. It was intended to fly non-stop across the United States and overseas at speeds much higher than the transport workhorse, the DC-4s (called C-54s by the military). Years later, I navigated the Connies from New York to London, Lisbon, Casablanca, and Liberia.

Pearl Harbor

I was washing my car when the news came. I was outside the apartment I shared with four other fellows on Westwood Boulevard near the UCLA campus. One of my housemates ran out and said to come inside, that there was a special bulletin on the radio. American ships in Hawaii had just been bombed by Japanese aircraft. There was massive destruction, and it was feared that many lives had been lost. Like it or not, the U.S. was now at war.

Though war had seemed inevitable for some time, there was still a certain shock and numbness once we were committed. I decided to get involved, even though—officially—as Lockheed employees we had draft deferments. This didn't last long. Soon after I left, a number of my colleagues were drafted.

President Franklin D. Roosevelt set a production goal of 50,000 military aircraft for 1942, our first year in World War II. It wasn't as impossible as it sounds. Most aircraft manufacturers had been perfecting mass production techniques to meet the British demand. Now the emphasis shifted from "we're doing it for the British" to "we're doing it for us."

The Constellation, known as the "Connie," developed for Howard Hughes' Trans-Western Airlines.

Japanese Internment Camps

There had been many warnings about possible Japanese invasion of California. One night, a few months after Pearl Harbor, air raid sirens began screeching their warnings. Defying all logic, I rushed outside to watch the searchlight beams converging overhead. They seemed to be following something high above the Los Angeles coast.

We never did find out what was happening. Was it just an attempt to alert the populace that we were at war? A false alarm? Or some sort of drill? In any case, the point was dramatically made.

Back in Turlock, Japanese citizens had been ordered to dispose of their properties quickly before internment in remote areas where they "would not be a threat." Actually, there did seem to be a potential danger—but not from the Japanese. You could visualize "patriots" taking out their fears, frustrations, and racism on the Japanese community.

My aunt, Esther Sward, was the local high school librarian. She was greatly impressed by the many Japanese students she had known and loved. After their farms had been sold for a fraction of their real worth, families were held in

Merced, California until the internment camps were completed. To show what an injustice was being done, Aunt Esther had my mother drive her to the nearby collection point so she could visit former students and bring them small bags of cookies.

A few months later, the California Senate held hearings in the community to gather evidence of Japanese subversion. Apparently they hoped to determine that the Japanese should never be allowed to return to California when the war was over. One witness, a former clergyman who had lived in Honolulu, testified that he had first-hand knowledge of Japanese "goings on" in the Islands. Others, perhaps with guilt feelings over their newly acquired properties, hoped his claims would prove true.

However, my aunt stood up in the hearings and naively expressed her feelings about her former students' scholarship, trustworthiness, good citizenship, and work ethic. She was soundly criticized by the local newspaper—"How could anyone in the school system dare speak out like that?"—and she nearly lost her job. Yet, no real subversive incidents could ever be confirmed.

I Grab a Protractor

Lockheed had been delivering their Hudson bombers to Britain for several years, using their own pilots. Now there was a rumor that navigators would be needed, and that military personnel would not be available. One evening in January of 1942, a fellow Lockheed employee named Lynn Sallee dropped by our apartment to have a couple of beers and say good-bye. He was going to TWA to be a navigation instructor. There was a man at the Van Nuys Airport named Charles Zweng, Lynn said, who claimed an association with P.V.H. Weems of Annapolis. Both were in the vanguard of aerial navigation studies.

Zweng's one room operation was called the "Pan American Navigation Service." Zweng was an elderly attorney and had published some manuals on the subject. His interest was entirely theoretical. I doubt that he had any practical experience.

Charles Zweng performed an unrecognized service by quickly supplying the books and gear essential to navigation training. He didn't teach personally, but without him, such training would have been unlikely. Not a penny of government money was spent on this training, and none was expected. Since I wanted to get on Lockheed's aircraft ferrying team, I forked over $150 of my own money for classes, another $50 for books, a calculator, a protractor, and an air almanac.

I couldn't attend night classes because of my job, so the extent of my celestial experience was daytime observations of the sun and moon. There was no beginning or end to these classes. You just jumped in and learned whatever component was being taught. Sooner or later it all made sense.

One day a Colonel visited the Van Nuys airport classroom and told our instructor how tanks were getting lost in the North African desert. Would a better knowledge of navigation be helpful? I doubt if that project ever developed, but we were impressed that someone else thought we might be on to something that could prove useful in the war.

But by May of 1942, none of the students had been able to get a navigation job. I was about to write off my effort as general education when some one showed me an ad in the Los Angeles Times. Pan American Airways was hiring in all job categories. I immediately contacted a Mr. Ramsey who said I was the first navigation applicant who had "experience," and that they would probably be able to use me.

It took several weeks to get released from Lockheed after Pan Am made me a definite offer. The pay at Lockheed amounted to about $500 per month with overtime. I had been able to save $200 of that in War Bonds. Pan American offered $165 a month until I checked out for flight. Then my pay would go to $195 per month, with $1.25 a day for laundry and tips when overseas.

I was now a navigator.

CHAPTER 3

Navigation School

*Pan American Airways' Pacific Division
is taken over by the Navy and begins
intensive training of pilots, radio
operators, flight engineers, and
navigators at Treasure Island.*

On July 5,1942, I was among about twenty who
reported to Treasure Island in San Francisco Bay for navigation
training. The classes were held in Pan American's Pacific
Division headquarters, formerly the tower of the administration
building for the 1939-40 Golden Gate International Exposition.
This huge fair, that had looked forward so hopefully to a world
of peace, was now the site of our wartime training.

I confess that my main thought was that navigating
would be a good way to participate in the war without being a
foot soldier. It never occurred to me that, at twenty-three and
without a military background, I might become an officer.

The program was a repeat of what I had already done,
but most of the other men, recruited from Southern California
colleges, had no previous navigation training.

*Before my first flight,
checking my charts on
the floor of my mother's
living room.*

From All Walks

A classmate, Herb Fahnestock, came to the program with unique qualifications. His family was well known on Wall Street. When the Pan American personnel interviewer asked how he had become familiar with navigation. Herb replied, "Oh, on the family yacht."

The Fahnestocks had interests in Shanghai Power and Electric which had been confiscated by the Japanese when they occupied the city. Herb's parents happened to be in Manila at precisely the wrong time, and they were imprisoned there in the Santo Tomas concentration camp. Herb had immediately tried to get a commission in the Navy, but was rejected because he was too slight of build. Of course, he fit right into the Pan American program. He was later commissioned with the rest of us.

Herb told us he'd learned that someone close to his family and still in Shanghai during the Occupation was playing dirty tricks on the invaders. This man was following the Japanese meter readers around, disconnecting the meters as soon as they had been read, and then reconnecting them just before the next readings. This small gesture of defiance gave the users free electricity and cost the power company a bundle.

Herb's mother died in the concentration camp. Understandably, he felt very bitter toward the Japanese for many years. In 1993, our navigation group was having a reunion in Honolulu and decided on the spur of the moment to stop at a Japanese restaurant. Afterwards, Herb's wife confided that it had taken over fifty years for him to do anything Japanese-related. But he was too much a gentleman to protest what the party wanted to do.

Our Little Red Cub

One of the attractions at the 1939-40 San Francisco World's Fair on Treasure Island was the Pan American Clipper base. From there, the giant seaplanes made their weekly departures to the Orient. Visitors were allowed to view the planes in the hangar as they were made ready for their next trips.

To demonstrate the size of the Clippers, a small Piper J4 Cub Coupe was parked under its giant wing. This little plane was a carefully polished, fire-engine red, with wheel "pants" to help streamline the landing gear. When the fair was over, the plane was going to be used by a Pan American employees flying club. However, the attack on Pearl Harbor generated a ban on private flying in Pacific coastal areas.

Shortly after we finished navigation school, four of us chipped in to buy this Piper Cub for $1000. Pan Am was still hiring pilots with 200 hours flight time. We now had our very own plane for training. Maybe we could get the commercial license and transfer to the pilot group.

Somehow we got it to Nevada where it could be flown. Amazingly, I never asked how he did it! We found "tie-down space" at a dirt field in Lovelock, Nevada, which became our training base. There was a small flight training school for military-bound pilots. Our plan was to get off-duty instructors to give us training in our own airplane. We would fly into Oakland from Hawaii exhausted, grab some sleep, take a night train to

Nevada, and try to find someone to give us flight training the next day. Of course, the prime time was reserved for military students. What time was left was usually windy or too near darkness to be useful for training. In a week of standing by, you might log one or two hours of instruction. It took a number of trips to Nevada over many weeks to get the eight hours of solo flying so we could fly without an instructor.

Our little red Piper Cub "27009"—which somehow got to Nevada.

Gasoline was rationed but plentiful for flight instruction which presumably would contribute to the war effort. After I got my private license, I took long cross-country trips to log time for the magic 200 hour requirement. A two-day trip to the Seattle area was followed by trips to Ogden, Utah; Cheyenne, Wyoming; Dodge City, Kansas; and to Tulsa.

In early aviation, railroad tracks were referred to as "the iron compass." To return to Nevada, we sometimes used them, but mostly we followed Route 66. It was a bit nostalgic retracing my childhood journey.

One time, I was eager to get to Las Vegas, so I flew north of Winslow, Arizona to intersect the Colorado River and followed it to Lake Mead and the Grand Canyon area. This was a very stupid thing to do. First of all, there's a rule with small aircraft that you don't fly over any place where you can't land in an emergency. And you definitely don't fly a tiny airplane in heat and turbulence of summer over the Grand Canyon. If I'd gone down, they'd never have found the body. But I had a date that night in Las Vegas and didn't want to miss it.

By this time it was September of 1944. Pan American was not hiring. Even if they had been, they could have gotten far more qualified ex-military pilots. Wiley Umstead, my co-owner of the airplane—by this time, the other two had dropped out—got his commercial pilots license, was later accepted as a pilot for Overseas National. He eventually became a 727 jet check-pilot for Japan Airlines, authorized by the government to put other pilots through their paces for their annual checkup.

Boats with Wings

International commerce has traditionally been between two ports—seaports or river ports. Virtually no major cities have grown up without a port. When the airplane became the new means of travel, it used these same cities as destinations because that's where the commerce was.

There were two reasons that a sea plane was the logical way to go: Landing gear for heavy aircraft had not been developed, and departure and landing points were always near water. Using a "boat" also made passengers feel more secure about flying over water. It was, after all, a terrifyingly new experience. All those thousands of miles of open ocean and nothing between the passenger and the sharks except the little-understood laws of aerodynamics. The theory was that the navigator would continue navigating the craft if it had an emergency water landing. If by chance the ship sank, everyone could scramble into inflatable rubber life rafts and keep going. I felt this was wishful thinking, but that was the official emergency plan.

"...even at this distance, I thrill to the wonder of it all," said President Franklin D. Roosevelt from Washington, D.C. on the inaugural flight of the China Clipper in 1935. (Before there was an Airforce 1 plane, President Franklin D. Roosevelt was secretly flown overseas by a Pan American B314 seaplane to meet with Churchill and Stalin.)

The Navy Takes Over Pan Am in the Pacific

Before Pearl Harbor, Pan American Airways was the only way to fly across the Pacific. Now the U.S. Navy needed to get equipment to Hawaii, and they had no long-range seaplanes capable of carrying substantial loads. It was logical for them to acquire Pan Am's whole facility. Overnight, Treasure Island became part of the Naval Air Transport Service

The plan was to use Martin seaplanes, generically called China Clippers, and Boeing 314s to fly between Treasure Island and Pearl Harbor. This 2,400 statute-mile hop was the longest currently being flown in the world. The Navy quickly realized that they now had the resources to extend flights from Hawaii to Australia where Douglas MacArthur was currently head-quartered. This could be done by "island hopping" across the Pacific, with the longest hop about 1,000 miles.

So the Navy gave Pan Am some PBM two-engine patrol bombers. To man these planes, they had to abandon their slow-and-easy peace-time apprenticeship training. Before the war, Pan Am had been hiring young pilots and cross-training them as navigators, radio operators, and flight engineers over several years. The idea was that Pan Am's future flight captains would have experience in both flying and ground operations. The senior pilots would become "Masters of Over-Ocean Flying," much like captains on ships who acted as gracious hosts while being available to consult with the working crew.

Nobody had any idea that ocean flying would accelerate so dramatically. Now, they had to break out of its "slow speed, safety, and good public relations" mode. With the world in crisis, everything had to accelerate. All categories of flight personnel had to increase far faster than the old apprentice program would allow.

Pan Am's first step was to make all the apprentices into pilots immediately. New trainees were recruited from young men with a minimum of 200 hours in light aircraft. Then Pan Am went to universities and hired mechanical engineering students, telling them "You're going to be flight engineers." They did the same thing to acquire navigators, hiring people like me. Radio operators were ex-Navy (retired) and Merchant Marine radio operators. Another group of recent college graduates were being trained as station managers for island duty.

Prior to Lindbergh's fight to Paris, there were no scheduled airlines. Aviators were not too well thought of, just nomadic daredevils. They flew by the seat of their pants because there was no instrumentation. To build public confidence, Pan American pilots were accorded the status of "supermen" by the press. A god-like arrogance implied that only Pan American could fly great distances with safety. To ensure a high caliber of personnel and weed out the uncommitted, some of the young junior pilots were subjected to humiliating hazing during their training. The rationalization was that it would make them better and more disciplined than the previous generation. I guess it worked. Soon, the Pacific Air Transport Command had many twenty-one-year old pilots flying newer, faster, and more productive equipment, and doing it without the grandiosity of their predecessors.

And so our intensive training began.

The PBM

The PBM Martin Patrol Bomber was used by NATS PAC (Naval Air Transport Service, PACific) between Honolulu and the south Pacific. It had a gull wing to keep the engines sufficiently high above the water, two Wright engines, and a magnificently strong hull. It flew at about 115 knots.

All armament was removed except for a pistol carried by the crew. Engine failures were not rare with this plane, so you had to be prepared to return to base or the closest facility on one engine. At island bases, the flight engineer became chief mechanic and worked around the clock with base personnel to do repairs. Often this required extreme ingenuity. For instance, they carried golf tees to push into holes in the hull if the rivets popped out.

A PBM (Patrol Boat, Martin) Mariner

The Philippine Clipper

The Philippine Clipper was the second M130 to be put in Pacific service, following the China Clipper. (There was only one China Clipper, although later the term became almost generic for any large plane crossing the Pacific). The Philippine Clipper was at Wake Island shortly after the Pearl Harbor attack, and was strafed by gunfire during the subsequent raid on Wake. The plane received only cosmetic damage, but the patches over bullet holes around the navigator's station were a constant reminder to us of the "incident."

This airplane crashed in a violent storm north of San Francisco in January, 1943, with Admiral Robert H. English and his staff aboard. Ordinarily, the aircraft would never have set out, but Admiral English was trying to deliver information that was considered vital to the war effort. The wind velocity in this storm was close to the air speed of the plane which made for erroneous assumptions of position when their navigator tried to depend on the quickly shifting radio bearings. Normally, a quick shift would mean that the plane was close to its destination, but not when your speed is supposed to be 90 mph and you're in a 90 mph wind. Your speed becomes 180 mph. Thus, the plane was way north of San Francisco, and didn't realize it. But if it had turned around and tried to fly into a 90 mph wind, it wouldn't have gotten anywhere. The plane simply should not have been in that environment.

Consolidated Aircraft PB-2Y3, the Coronado.

PB-2Y3 Coronados

Later in the war, we got the faster PB-2Y3 Coronado four-engine Patrol aircraft that flew around 130 knots. They had a longer range and could carry bigger loads. The Coronados supplemented the four-engine Clippers between Honolulu and San Francisco. To carry maximum loads eastbound from Honolulu, the distance was sometimes shortened fifteen miles by making a short flight to Kaneohe Naval Air Station at Kailua to refuel. This made for a longer day, but, theoretically, it contributed to safety. Once we began island-hopping toward Australia, the greater speed of the Coronado meant we could overfly islands like Palmyra, cutting several days off of a round trip.

CHAPTER 4

How to Navigate

> *We learn the basics of navigation, first
> at Treasure Island and then in the
> cauldron of wartime flights across the
> Pacific.*

We soon learned that navigating really wasn't hard. I was an experienced draftsman, and navigating is just drafting on a piece of paper 2,000 miles wide. But there were other skills I needed to learn.

Celestial Navigation

In marine navigation, the distance a ship has traveled between the time a navigator takes a celestial fix and completes his computation is negligible. The ship is traveling only eight to fifteen knots, and accuracy would be within a few hundred yards. However, in an aircraft, we hoped for an accuracy of three to five miles. The navigator tried to complete his sighting and plotting within fifteen minutes so that course corrections could be made while still near the fix. One very bright trainee kept trying to be exceptionally accurate at the cost of being too slow. A good fix quickly done was better than 100 percent accuracy too late to be effective.

You had to be able to identify twenty to thirty stars to do celestial navigation. The constellations provided easily identifiable patterns for major stars, with star maps helping to identify those used less frequently. Ideally, a star dead ahead or behind the plane was chosen to determine progress. Another one "abeam" to the left or right was picked to show how close to course the aircraft was. A third star in between the other two was used to confirm the accuracy of the other two. The navigator could view the stars from a bubble-like window in a crew room aft of the actual operating stations. This window could be opened to observe the fainter stars. Overhead cloud cover could obscure the stars, interfering with timely observations. This was always a cause for concern.

To "plot a fix," you recorded ten observations for each of the three stars—the dead-ahead, abeam, and reference stars—in three two-minute periods. Then you used the average for your calculations. Sextants can't be used in an airplane. They require a steady horizon to measure the angle of elevation for a celestial body. Instead, we made observations with an octant that had a bull's-eye level similar to that in a carpenter's level. (Better octants were developed later in the war, with averaging devices that made for improved accuracy.)

The aircraft had to be held steady during each of the three two-minute celestial sightings. Some pilots understood and cooperated. Some didn't. Any change in airspeed or attitude would shift the bubble, resulting in less accurate readings. Rough air and cloud tops kept the job interesting. The whole process took about twenty minutes if the skies were clear and the plane was steady.

Ideally, these celestial observations were done every two hours. In between fixes, the navigator would toss a magnesium flare overboard. In the daytime, its smoke would indicate the direction and force of the surface winds. At night, the glow let you measure the angular drift using a pelorus. To reduce weight, only a dozen flares were carried. They were rationed with great care. You might have a critical need for them later.

At altitudes of 4,000 to 8,000 feet, middle and high clouds often obscured your view of the stars. You might wait as long as fifteen minutes for a break in the overcast just to get a peek at a star. Then you'd observe it with the octant, do minimum averaging, and hope the data worked out. These observations were done using a flashlight. With the star-sighting windows open, it was cold!

Of course, in the daytime the sun was the only celestial body we could count on. There were times when the moon, and Venus were visible, enabling a celestial fix in the daytime. To observe Venus during the day it was necessary to calculate its direction and angular altitude, preset the octant, turn it to the proper direction, and start the search. Sometimes I felt I was shooting my imagination rather than Venus; but it did seem to work.

Sometimes a very bright light would appear to the west just after sunset. This was often thought to be an enemy aircraft due to its unusual brilliance. Of course it was Venus which would eventually follow the sun into the sea.

At equatorial latitudes where the sun was directly overhead and near the latitude of our position, a "noon" fix could be determined by observing the sun just before it passed our longitude, a second time as it was due north or south of us, and a third time after the sun was west. These quick changes of azimuth (direction of the sun) gave us three or more lines of position with only one celestial body. But this was a once a day opportunity. There was a graphic way to do these calculations which I always enjoyed in equatorial areas.

Chronometer, Clock, and Compass

To know where you are at sea, you need to know your latitude (the lines on a globe that parallel the equator) and your longitude (the lines on a globe that connect the north and south poles.) Latitude has always been easy, but, until the eighteenth century chronometer, longitude was pure speculation. Without knowing both longitude and latitude, navigation was mostly an educated guess.

It's possible to determine latitude without a chronometer because latitude isn't a function of time. You can calculate your latitude in the northern hemisphere whenever the North star is visible. Sailing ships navigated to the same latitude as their destination, turned west (or east), and sailed on the latitude line to their destination. Historians tell us that pirates established themselves on a popular latitude line and awaited the arrival of vessels to be plundered. The Polynesians used the same principles in their canoes for centuries.

A way to determine longitude, once a ship had lost sight of land, was one of greatest searches in scientific history. For centuries, astronomers, mathematicians, and inventors, from Galileo to Newton had attempted to solve the problem. In the eighteenth century, the British Parliament offered a reward of £20,000 ($12 million in today's money) to anyone who could come up with a practical way of computing longitude at sea.

Clockmaker John Harrison earned the prize with his chronometer, an extraordinarily accurate clock "that would carry the true time from the home port, like an eternal flame, to any remote corner of the world." One of the thorniest dilemmas for navigators had been solved.

Chronometer accuracy was paramount. Shortly after takeoff and before any celestial observations were needed, the radio officer would obtain an observatory "time tick" to confirm the accuracy of the plane's chronometer and our watches. In lower latitudes, an error of one minute could create an east-west error of up to fifteen miles, which would generate a ten minute error in arrival time (at an airspeed of 90 knots), or a fifteen-mile course error in north-south travel.

After the plane leveled off at flight altitude, the navigator would determine the compass error. It was different for every plane, and could be affected by cargo aboard. These errors could be five degrees or more. An accurate compass was paramount in case there was any subsequent difficulty in obtaining fixes.

Fred Noonan and Amelia Earhart

Pan American's first navigator back in 1935 had been a man named Fred Noonan. I was told by those who knew him that, being first, he did more or less as he pleased. He was lax about taking deviation checks on the compasses, a carelessness that would be disastrous if you were depending on dead reckoning alone for a landfall. To start a long flight with undetected compass error is pure folly. We always figured that was what happened to Amelia Earhart who vanished during her round the world flight in 1937. Fred Noonan was her navigator.

Earhart and Noonan had another strike against them. Prior to World War II, there were hundreds of islands in the south Pacific which had no economic value to the rest of the world. Pilots were still flying by marine charts, and these charts were plotted so ships could avoid running into islands, not because anyone wanted to find them. Some were as much as twenty miles from where they were depicted on aeronautical charts. I wonder if Fred Noonan checked the actual position of Howland Island, Earhart's destination when they disappeared.

Weight and Fuel Consumption

When Charles Lindbergh flew solo over the Atlantic in 1927, weight was a major consideration. He even trimmed the borders off his navigation charts to save a few ounces. This was still true fifteen years later on the San Francisco to Honolulu flights. Anything not absolutely needed was left behind, so more fuel or a bigger payload or an extra passenger could be carried. The clothing and gear we needed for our stay in Honolulu or the South Pacific were left in the local hotel, to be used when we returned. That way, we walked on the plane with minimum luggage. (I sometimes think nostalgically of those days when I'm waiting at a modern-day baggage carousel.)

When large seaplanes took off heavily loaded for a sixteen to twenty-two hour flight, they couldn't climb until they had burned off some fuel. It was common to remain at 1,000 feet before climbing in stages to a final cruising altitude of 6,000 to

9,000 feet. Not very high by later standards. Cockpit windshields were often coated with a thin layer of salt from the sea spray.

Prior to departure, the Pan Am dispatcher would give the captain and navigator a flight plan which divided the route to be flown into segments, usually 5 degrees of longitude. The plan took advantage of the best meteorological winds available at the appropriate altitudes. True courses and headings for each segment were calculated. The summation of time for each of these zones would give the total time and fuel needed. Extra fuel reserves were added, allowing for the need to "hold" over a destination and then to fly to an alternate where weather conditions might be more friendly.

The navigator then confirmed the accuracy of the flight plan, using the meteorological charts. The captain would approve the total, possibly requesting extra fuel as insurance. We felt this was a nice gesture—it could mean flying with fewer passengers or cargo, but it increased safety in case of bad weather or an engine shutdown. If an engine conked out, the remaining engines had to carry the burden at a lower airspeed, and this increased fuel consumption.

There was a common impression, never put into words, that pilots and flight engineers arranged for "pocket fuel" over and above what was needed for the flight—so many extra gallons for each wife and child of crew members—as a sort of insurance policy. However, carrying extra fuel used up fuel just to transport the extra weight, lowering the performance of the aircraft. So it was a constant Catch-22 situation.

A rule of thumb was that half the extra fuel would be consumed just to carry it to the plane's destination. The military fuel cost twelve cents a gallon. Obviously, deciding on the exact amount of fuel to carry was a weighty process.

Fuel Dumping

Airplanes lose weight as they fly, burning up their heavy load of fuel. Fortunately, simple physics dictates that they can take off with a heavier load than they should have when they land. Takeoffs are usually fairly smooth, but the frequent bumps on landing can damage landing gear and hull or put heavily-loaded wings under stress.

If you have to land the plane before it has burned off enough fuel to reach an acceptable landing weight, you usually dump some of your fuel down special chutes—over water and open land only, never over cities. This creates a long vapor trail aft of the plane.

Fuel dumping is safe in today's faster airplanes, but, at the slow speeds of the early flying boats, it presented a danger. It was theoretically possible for lightning to strike the vapor trail and ignite the gaseous mixture. If the flame traveled faster than the plane's airspeed, it could catch up --- with disastrous results. All this may sound highly unlikely, but some believe this is what happened to a Sikorsky Clipper on one of the first proving flights in the south Pacific.

Ironically, full fuel tanks are safer than empty ones which are full of potentially explosive vapors. It was standard practice to refuel as soon as possible after landing. This had the double advantage of avoiding empty-tank hazards and making quick, unscheduled departures possible.

Wind and Weather

About 1940, a Clipper headed to Honolulu hit several hours of rough weather, and the navigator was unable to observe the stars for a position fix. They radioed their problem to the San Francisco Flight Watch. A man named Albert Francis, a meteorologist with Pan American Pacific Division, was awakened at home and asked for his advice. Francis reasoned that the plane was flying in and parallel to a weather front. In other words, it was traveling along inside of a moving front, as if it were wearing it for an overcoat. If the aircraft would turn to the right or left, said Francis, they'd fly out of the front and into better weather. The plane did this, the weather soon cleared, and the navigator could observe the stars once again. (Albert Francis had been meteorologist on the ill-fated Navy dirigible Macon which crashed off the coast of California in 1935 with eighty-three casualties.)

In those days, there were no satellites to provide weather pictures. Data was received from other aircraft or even surface vessels. Even in 1942, weather data might be several days old because no one else had recently flown the critical leg between San Francisco and o Honolulu.

Flight engineers constantly monitored their equipment for the latest weather news, so they could pinpoint any problems needing correction before landing. Crews relied on their own ingenuity. The radio operator reported the position, mechanical status, observed weather, and fuel remaining every hour. These fellows also supplied us with a constant flow of information, including sports scores and tomorrow's news. They got information before the newspapers had it and often passed it on to the passengers, a bonus service.

Meteorology was important for choosing the most efficient long-range routes. We learned that the shortest flying time isn't necessarily a straight line to your destination. By deviating off course, you can go around weather systems and enjoy more favorable tail winds.

Each seaplane flight from San Francisco to Honolulu was a challenging game for the navigator. The aircraft was the chess piece. The Pacific was our checkerboard. To travel the 2,092 nautical miles at 95 knots in still air would take twenty-two hours. To carry any substantial load in less than twenty-two hours meant finding a tail wind and a route that would provide it. The meteorologists gave their best estimates, but it was up to the navigator to make the right moves in this game with nature. When the weather was bad enough, aircraft couldn't make the trip with enough payload to justify the flight. The other crew members were probably unaware of the vagaries of air circulation, so the navigator was very much alone in this critical contest.

To succeed, we had to get the plane and its load where it was going on time. As exhausting as it was, I always got great satisfaction from playing and winning this game with the elements. Today, jet aircraft airspeeds are so much greater than probable winds that adverse winds are not critical.

Measuring Drift

If the wind was moving the aircraft to the left, the navigator could steer into the wind, using a correction angle equal to the drift. This would keep the plane on the intended course. But if the wind shift was expected to cause a later drift to the right, then the aircraft would be back on course without a penalizing correction being made, and with a possible greater tail wind midway in the process. At an airspeed of one hundred knots, a head wind or cross wind of thirty or forty knots used up quite a bit of your forward airspeed. Fortunately, we didn't have to worry about other air traffic so we could take full advantage of "wind circulations."

The navigation table where I worked was about four feet wide, with a drafting machine attached. Aft of the table was an optical drift meter that could be pushed outside the plane to measure the angle of drift caused by the wind. It consisted of a periscope with parallel lines inscribed on the viewing surface.

When white caps on the sea surface were visible, you could turn the parallel lines of the drift meter so the surface traveled along this grid. The angle needed to accomplish this was the angle of correction necessary to compensate for wind drift error. But if there were clouds beneath the aircraft or the sea was slick, the drift meter was useless.

When this happened, you could toss a small magnesium flare overboard. Then, using a pelorus (a device for measuring the relationship of observed objects in degrees) you'd measure the angle from the flare. On the M130 (the China and Philippine Clippers), the pelorus had to be mounted on a passenger's window sill. At night, it was necessary to wake a sleeping passenger, install the pelorus in an opened window, throw out the flare, and wait for the light or smoke to be visible. If you couldn't see it, you had installed the pelorus on the wrong side of the plane. You had to move the instrument quickly to the other side, wake up another passenger, and hope to get your information while the flare was still visible.

However, periscopic drift sights were rarely adequate. So, on the PBM seaplane, we would go below into the hull, crank open the hatch covering the bombardier's window, and fashion a crude drift sight using a protractor and string. Usually, the navigator could determine the surface wind direction by streaks in the water, and come up with a fairly accurate wind velocity by viewing the condition of the sea. This would be fairly obvious. A glassy sea meant no wind. The more white caps in the sea, the stronger the wind.

When flying at 1,000 feet to 4,000 feet, we could come close to reckoning the wind at our altitude. Occasionally, this could be determined by a "double drift" observation, using three flares and changing the aircraft direction forty-five degrees, then ninety degrees in the opposite direction, then forty-five degrees back to the original heading, with drift being observed on each heading. This maneuver took three valuable flares, and reduced the forward speed to seven-tenths of normal.

Navigation was a matter of judgment, experience, and good luck with the weather. A reckoned position could be off by fifty to one hundred miles if the airspeed was ninety to one hundred knots, and a radical wind shift occurred between celestial fixes, or no fix had been possible. There was no one else in the air, so a collision was not likely, but forward progress ninety miles shy of where one hoped to be meant an extra hour flying with remaining fuel a factor.

Dead Reckoning

There were no radio ranges at sea for guidance. In the daytime, there were no stars, so accurate fixing of position was unlikely unless you had a visual sighting of a reef on the National Geographic maps. Under these conditions we were said to be "dead reckoning." This meant that we deduced our position by reasoning that our air speed altered by wind gave us effective

Boeing 314 Clipper flight deck.

velocity over the sea, while our compass, corrected for geographic variation and known compass errors due to a particular air craft and its cargo, gave us direction...provided that a wind drift correction was included. When you combine speed and direction, you assume your most probable position. And you keep repeating the process with updated corrections—sightings or celestial fixes. This is the basis of navigation.

Using Sun Lines

A variation of the sun line approach was used by mariners for the several centuries between Columbus and the invention of the chronometer in the 1700s. To locate a small coral island in the daytime without radio aids, we used a similar method. It was a last-resort for making landfall if no fixes had been obtained, and the navigator felt that the "dead reckoning" was suspect. I liked to practice it now and then, just to see if it really worked.

Cornelius Dunbar, who had been the first Air Corps pilot accepted into the military training program without a college degree, was very interested in this celestial trick. It could only be practiced when you could get extra cockpit cooperation. Some pilots were reluctant to bother practicing these sun line approaches.

The procedure was this: Plot a sun line, pick it up with parallel rules, and move it over the island you are trying to reach. Then make a direction change away from the island. Navigate to that line, arriving at it with enough clearance so you know the destination is either to the right or left. Make the appropriate turn, and start letting down. The trick is to be far enough off course that you don't turn the wrong way. Keep taking sun lines and correcting to fine-tune the track accuracy.

This was the navigator's show, and you could be a real goat if you made an error. The estimated time of arrival (ETA) was always compared to the actual arrival time. It could be a source of embarrassment if the navigator was off too much. Of course, we could always claim there had been a wind shift or change in air speed.

Sight Reduction Tables

The "sight reduction tables" had originally been designed for marine work and ships. When you observe the stars, you also note the time, your approximate geographic location, and their angle above the horizon. Then you have to refer to the publications of the Hydrographic Office to make your calculations. This laborious process was okay when you were in a slow-moving ship, but not in an aircraft where time was of the essence.

Parachutes

In an attempt to protect passengers, the government considered requiring parachutes on board. This was not a good idea for obvious reasons. Parachuting into frigid, shark-filled water is much more dangerous than sticking with the plane which would probably float. The weight of the parachutes would be significant, increasing the danger, and in the end they would probably be useless.

Before the war, Pan Am had tried to build public confidence by advertising that their plane was as good as a boat. However, neither the Boeing 314 nor the famed Martin 130 China Clipper had pontoons outboard below the wings for lateral stability on the water. The earliest Clippers had "sponsons," stubby wings that were supposed to perform like pontoons. But they couldn't handle a cross wind on the water without risking a wing dipping into the sea. This had happened several times. So I didn't believe that a forced landing in the ocean was quite as safe as advertised.

Reporting Position

Radio communication was by Morse code with transmission by high speed "bugs" operated by ex-Navy or Merchant Marine radio men. We radioed our position report every hour at an assigned time. This included weather and sea conditions. If our report was late, Flight Watch would assume there was a problem and trigger an alert. Avoiding a false alert kept the navigators busy. We had to calculate the "best position" (where you thought you were), fill out the report, and "make schedule" (get your report in on time). Then the radio officer on the aircraft had to code the transmission for security reasons.

Takeoffs

A heavily loaded seaplane is really still a boat when power is first applied for takeoff. As it speeds up, it rises slightly and skips along the surface. Midway along the bottom of the hull is a sort of notch. It breaks up the surface of the water and allows the plane to come free of the "suction" holding it to the surface.

Whenever the surface wind was calm and the water glassy smooth, there were times when the plane could not break the surface and become airborne before the engines overheated. A choppy water surface was actually a real plus for takeoff. A second try after a failure on a glassy surface was usually successful because the first try had stirred up the water.

Sometimes a motor launch was sent out to disturb the surface of the water.

It could take as long as forty-five minutes between leaving the dock or anchorage and becoming airborne. Floating debris and other ships or boats in the area could make for difficult takeoffs, requiring long stretches of San Francisco Bay.

Wind was the Clipper's enemy. A really high wind could break the anchoring line and blow the plane away, both in a coral lagoon or at home base on Treasure Island. Even a brisk cross wind could dip a flying boat's wings into the water. The only solution was to keep the engines running and have a flight engineer remain aboard to keep the aircraft headed into the wind. Sometimes he had to do this all night.

Checking Out the Aircraft

Extensive safety drills were held to ensure that everyone knew exactly where all emergency gear was located. Each crew member had specific assignments in case the aircraft went down at sea. Any smoke in an aircraft would mean that the crew could be virtually blind. That's why it was essential to know where everything was in case of an emergency.

An hour before departure from San Francisco, crews reported to Mr. Sales, a retired Navy Chief Petty Officer, to be tested in seamanship and emergency equipment drills. Each crew member was given a stack of poker chips individually labeled "fire extinguisher," "life raft," "axe," "Very pistol flares," "water canister," "escape door," "Mae West" (radio transmitter), etc. We had to position each chip on a model drawing of the airplane, indicating where these objects were stowed. Then, everyone had to recite his emergency assignment with precision or face being removed from the trip. To be replaced because you didn't know the positions of your emergency gear would be more than embarrassing.

In case the plane ditched in the sea, the navigator's assignment was to gather up all his navigation gear and make certain it got aboard his assigned life raft. The prospect of jumping into a rubber raft carrying sharp-pointed drafting dividers seemed rather self defeating to me. I think the idea that we were going to sail rubber life rafts anywhere was mostly a symbol of not giving up.

At Treasure Island in San Francisco Bay, each crew member went aboard the aircraft to make certain his gear was in place and operating properly. The navigator checked his charts, octants (two), drafting gear, pelorus, magnesium smoke flares, etcetera.

Later on in the war, we had to take special care to check out the inflatable life jackets before departing certain island stations. They contained small cartridges of compressed carbon dioxide which, when punctured, inflated the life jacket. Unfortunately, someone discovered that the cartridge could also be used to produce carbonated drinks for island personnel. This clever misappropriation could have proved life-threatening in an emergency.

The crew then assembled at Operations and marched two-abreast to the boarding dock where the Captain received the ship's papers from the station manager before boarding. I had observed this pageantry, done for the benefit of the visitors to the 1939 World's Fair. The departure of a Clipper for the Orient was special, and the sight of twelve to fifteen men marching to the Clipper late in the afternoon drew crowds. Now I was one of the marchers.

We were instructed not to carry our bags aboard, but to leave them in a specific place where they would be boarded by baggage personnel. This was fine in San Francisco, but the PBMs flying south out of Pearl Harbor were strictly a military operation. The first time I left my bag as per instructions, expecting it to be loaded, it wasn't. So, for the next ten days, I wore the same clothes as we flew to Brisbane via Palmyra, Canton, Fiji, and New Caledonia, and back the same way, with one flight per day. When I reached Pearl Harbor, my bag was right where I had left it. I learned something.

Point of Decision

If the aircraft was deviating from the flight plan, and the trend indicated you didn't have enough reserve fuel to reach your destination, then there was a point beyond which you must not continue. This "point of decision" took into account the possible loss or malfunction of an engine, which would mean greater fuel consumption per mile. We knew what conditions we'd just flown through, but conditions ahead were estimates unless we got reports from another plane that had crossed the area.

"Equatime" meant you had reached the point where the flight time to your destination was the same as that to your point of departure. A celestial positioning before reaching that point was essential. Should you proceed? Or should you turn around and go back? There were times, especially in the winter, when you'd spend twenty-plus hours in the air and end up back where you started.

Floating Buoys

A seaplane in the water is a boat, subject to the current and the wind. Maneuvering it alongside this anchorage was awkward. Before our flying boat could be secured on the surface, we had to come abeam a floating buoy, retrieve a line with a loop, and quickly slip it over a bow post. The engines had to be operable until the exact instant the bow line was secure on the bow post of the plane. Then they had to shut down immediately. When it was time to reverse the process, casting off had to be done at the instant both engines were operating. If only one engine was operating, the seaplane would go in a circle. If neither was operating, the aircraft would drift without being under control. As the navigator, I often got this assignment in the bow. It was very tricky to spot the floating buoy, grapple for the line, and attach it quickly to the bow post.

On the first flights into Hong Kong, an enterprising Chinese boat woman known as "Mary" decided she could be of assistance at the buoy. She was accurate, dependable, and friendly, so pilots came to rely on her. Her compensation was permission to have the garbage from the flight.

How the Polynesians Did It

The Polynesians were navigating the Pacific Ocean from New Zealand to Hawaii for centuries. They guided their canoes using their knowledge of star movements, color of sea water, tidal variations, oceanic swells, and even the presence and actions of birds. Each variety of bird has a different flight radius from its home base. If you spot a bird that can fly only fifty miles from its nesting area, the navigator knows he is within fifty miles from landfall. By observing the various bird species, they could gather a lot of information on their positions.

Clouds form over islands, formed by the heat of the land. Low-lying islands cannot be seen from a distance, but the clouds can.

A sextant can be made from a coconut shell although it can measure only one angle. If holes are made around the circumference and it is then filled with water, the water's surface becomes a level or artificial horizon. A notch is cut on one edge to form the number of degrees wanted for the particular situation, for example twenty-one degrees for Hawaii. When you are on an island in the northern hemisphere, the angle from water level to the North Star, Polaris, is within forty-nine nautical miles of the latitude. At the North Pole, Polaris is overhead, roughly ninety degrees from the water level. On the equator, Polaris' altitude or angular measurement, if seen, would be zero which is the latitude of the equator. When the correct time is known, the errors can be corrected.

The Polynesian navigators, who had no time-measuring equipment, could sail north toward the Hawaiian Islands. They observed Polaris at night until the star fit in the notch, indicating they were at the same latitude as their intended landfall in Hawaii. If they were south of Hawaii, Polaris would be below the notch in the shell. If they were north of Hawaii, Polaris would be above the notch. When they reached the correct latitude, they had to guess whether they were east or west of the island. To resolve this, they would travel either east or west, watching for birds or clouds. If they didn't find Hawaii (heaven forbid!) they'd turn around and go the other way until they ran into it. These same simple methods, greatly refined, were used by modern navigators in emergency during the seaplane operations in the south Pacific areas.

Only a few Polynesians are now trying to preserve this knowledge of the sea and stars.

"Look Out the Window!"

When I started with Pan Am, I assumed that I'd be working with sophisticated equipment. However, navigation was still pretty primitive. We hoped we'd be accurate at the end of the journey, but, until then, we could be anywhere. Fortunately, we weren't going to collide with another aircraft. There was no competition. However, it was essential to compare our progress to the flight plan.

Each south Pacific trip was a game and challenge. If we could spot a reef or small coral island en route, it gave us positive and effortless positioning. I recall one small reef named "Curondelet" that was not always above water level. It was probably much less than an acre, usually visible only as white choppy water. By referring to the National Geographic maps and Hydrographic Office publications, I'd know where we were. The accuracy and quantity of their information showed years of research and data accumulation. Such visual sightings were a bonus and were easier and quicker than celestial sightings.

Those of us who qualified as navigators first were assigned students to train for a few trips until they were ready to navigate alone. It was standard to quiz the trainee during the flight to be sure it all made sense to him. We deliberately did this with a certain amount of pressure and urgency. We wanted to test whether the new man might panic if he ran into adverse conditions. When a decision had to be made without adequate data, would he use common sense?

I was checking out a student on the approach to Canton Island after a daylight flight. A few minutes before our estimated time of arrival, I asked him where we were. He had done a good job, all was well, and I expected him to stand up and look out the window.

Instead, he seemed to panic. He grabbed pencil and paper and started redoing his computations. He was probably afraid that if he looked and the island wasn't there, all was lost. So I had to reassure him: If you've done your work well, have confidence in it. Look out the window first.

CHAPTER 5

Taking Off

> *Navigating in the south Pacific for the Naval Air Transport Service in World War II.*

Throughout our training, we were told that we would also get some flight training around the San Francisco Bay Area before our first flights to Hawaii. Toward the end of ground school, we were fitted for a Navy-officer type uniform, and given passports and the necessary inoculations for south Pacific travel. Some of the PBM (Patrol Boat-Martin) aircraft, called Mariners, were already arriving at Treasure Island for conversion and pilot training. We were curious why we weren't getting familiarization training on the local flights which were being used for pilot and flight engineer training.

I was concerned not to have any flying time. I wondered if I'd experience airsickness when we hit turbulence. As a child, I had often been carsick, bouncing about in the back seat of the family automobile. It looked like my first airplane ride was going to be as navigator from San Francisco to Honolulu on the biggest seaplane in the world. What if I disgraced myself? But there were thousands of other fellows with less to look forward to, so I decided to think positively.

First Flight

My first flight as a navigator and my first flight ever were the same one. It was on the Boeing 314 Clipper, one of the most magnificent aircraft ever constructed, a seaplane with its hull built in compartments so that any damage could be localized. Its fuselage was divided into an upper deck for the crew and a lower deck for the passengers.

I remember asking before my first flight if I should bring my lunch. I needn't have worried. Meals were prepared in a galley by professional chefs and served on tables with fine china and silverware. The sofa-like seating converted to bunks for sleeping, in the same way that railroads converted their Pullman Cars at night. Clippers from Hawaii arrived at Treasure Island in the morning after breakfast. If the plane reached California earlier, the pilots were instructed to circle slowly at sea so that the breakfast service would not be rushed.

The stewards' duties were far from easy. They were all professionals from San Francisco hotels. Many of them kept their civilian jobs and worked them during their weeks off. At Island stopovers, the purser (head steward) had to replenish the plane's larder by shopping in local markets. This could be a colorful experience. On Fiji, for example, milk was delivered to the plane in whiskey bottles. We had a mix of military passengers, priority cargo, and mail.

The crew worked and slept on the upper deck, a spacious and comfortable environment compared to the cramped quarters of jet aircraft. A slow speed of 115 knots made for very long flights, so comfort was part of safety. The crew deck was close to the four reciprocating engines. For hours after landing, we could still "hear" the roar. However, that roar during a flight was much preferred to the dead silence of malfunctioning engines.

On the Boeing Clipper, there were hatches on both sides of the flight deck that the engineer used to access a catwalk inside the wing. He could go "aft" (in back of) of the engines during flight, and presumably make minor repairs while the

Just before my first flight in 1942, all decked out in my new uniform.

plane was in the air. A crew room behind the cockpit contained several cots for off-duty crew.

In 1942, our first year in the war, the China Clipper flew eighty-five trips between San Francisco and Honolulu—one round trip every four days. (One Urgent-Priority load turned out to be mop handles for some south Pacific island.) Later, when I was with Braniff International in the 1970s, their one Boeing 747 flew from Dallas to Honolulu daily, airlifting as much in one day as the China Clipper did in whole year.

Airplanes Don't Get Tired

One of my first profound realizations was that airplanes don't get tired. They are built to be in the air and don't benefit from being on the ground. They have to be fed, of course, and maintained. Many long distance routes were flown by planes that averaged eighteen to twenty hours a day in the air. A plane doesn't make money sitting on an airport tarmac, and flying doesn't abuse the equipment. The only limitations were on the stamina of the crews who flew them. Sometimes we would virtually live in the air for days.

Between San Francisco and Hawaii

On these day-long sea-plane flights, there were four or five pilots aboard. A normal watch was two hours, and the captain usually got a good night's sleep so he would be fresh for the approach and landing. Two flight engineers and two radio operators worked in alternate shifts, which made their work load reasonable. However, navigation was considered a one-man job with no relief at all.

A "great circle" routing from San Francisco to Hawaii usually meant making a slight course alteration about halfway there, at latitude 30 degrees north and 140 degrees west longitude. Traditionally, for some unknown reason, this was known as "Jones' Corner." Fortunately, most of the navigator's work had been done by the time the plane was 100 miles out from its destination. If radio facilities were available, the

Calculations for my first flight, 1942, from San Francisco to Hawaii.

navigator's duties were over. After his long hours of taking star shots and drifts, he would be exhausted, but it didn't matter.

Landfall on the Hawaiian Islands was usually associated with sighting the volcanoes atop Maui and Hawaii to the south. Several radio-range legs were receivable, and by this time it was daylight, so visual sightings took the place of star observations. Usually there was no time to go down for breakfast where the stewards were serving their second china-and-silver spreads.

For the return trip, one Pan Am captain had a really neat trick that made him the envy of his colleagues. He'd installed a two-way radio in his Berkeley home, very unusual for the time. When he flew in from Honolulu and saw the welcome-home sight of the Golden Gate Bridge was ahead, he would radio his wife to find out "What's for dinner?" and give her his arrival time at Treasure Island. She would be there to drive him home.

Scars in Eden

Before World War II, the Clippers landed adjacent to Ford Island in Pearl Harbor, then tied up at a facility in Pearl City. When we made our first flight there several months after the December 7, 1941 attack, we found considerable damage. The Gray Line buses that delivered us to Waikiki beach drove past a mile or so of destroyed buildings, many of them bulldozed into mountains of rubble.

At that time, Waikiki had only three major hotels: the Moana, the Royal Hawaiian, and the Halekulani. They were untouched, but even in the midst of all this beauty and relative luxury, there were stark reminders of the war. Every time we arrived at Pearl City, we were issued World War I gas masks to carry while we were in Honolulu. A machine gun emplacement stood on the beach next to the Moana. The fairways of the nearby golf course had been lined with old automobiles and other objects as a precaution against enemy aircraft landings. And at night, instead of strolling in the moonlight, we observed a complete blackout and a curfew. We could not leave the hotel after dark.

Pan American had reserved the entire sixth floor at the Moana Hotel for its personnel to use on layovers, paying six dollars a day for single rooms. Like all wooden buildings in Hawaii, it was riddled with termites. We joked that if the termites hadn't been holding hands, the hotel would have collapsed. (In 1993, twenty of our navigator group met for a three-day reunion at the massively renovated Moana Hotel. We got a special rate of $190 a day for the same rooms we'd occupied over fifty years before.)

Two large tables were reserved for our flight crews in the Moana dining room. We were given $1.25 a day for expenses. Food was free, but we were told to tip twenty-five cents for each meal. The remaining fifty cents was for laundry. One of the waiters, Steve, saved his tips. After the war, he returned to Manila and opened his own restaurant.

Down the beach, the Royal Hawaiian Hotel was occupied by Navy submarine crews on relief from patrol duty. In the afternoons, the hotel offered Big Band Music on the beach, often conducted by the top bandleaders like Artie Shaw. This was all for the benefit of the sub-mariners, but no one resented it. Those men deserved every luxury they got.

Drop Everything

It was my first flight from Honolulu back to San Francisco, and I purchased several flower leis for my mother as souvenirs. I was boarding the limousine, my bag in one hand, the leis in the other, when two passing sailors saluted me smartly. We did look something like Naval officers in our brand new uniforms, but I suspect the sailors knew exactly what they were doing. They must have enjoyed seeing me drop everything as I tried to return the salute.

"No Standing in Public"

Honolulu now seemed safe from further Japanese attacks, but it had another problem. Too many service men in a crowded "party" atmosphere seemed to invite interservice rivalry. That could end in brawls. The city lawmakers came up with an interesting way to limit the number of military personnel in bars. They made it "illegal to stand up in public while holding a mixed drink." That meant that, when all seats were occupied, guards at the entrance would not permit anyone inside until a vacant seat became available.

Hulas for Ohio

A weekly radio show, "Hawaii Calls," had been broadcast from the open-air lobby of the Moana Hotel for seven years. It was strictly for mainland consumption, and attracted many tourists who had been listening to it back home. We usually reached Honolulu after breakfast, and the show started at noon. That gave us time to shower and change into Hawaiian attire so we could relax in the trade wind breezes while watching the show.

Conductor Harry Owens had started the show in 1935. Their radio announcer was Webley Edwards. He became a well known politician after Hawaii achieved statehood. The singer, George Kainapau, had a truly sparkling smile, thanks to a small diamond that he'd had implanted in his front tooth!

This was the heyday for "native" Hawaiian music, much of which was composed by Harry Owens himself. In later years, the "Kodak Show" from nearby Kapiolani Park replaced "Hawaii Calls" at the Moana.

The Dole Race

After Charles Lindbergh's New York to Paris nonstop flight on May 21, 1927, other pilots rushed to make equally spectacular flights over different routes. They frequently had commercial or private sponsors, which generated a lot of publicity for both flyer and benefactor. One such booster was James D. Dole, a Hawaiian pineapple grower, packer, and aviation enthusiast. He offered a $25,000 prize for the first person to fly nonstop from California to Hawaii within one year after August 12, 1927. (The word "nonstop" in this context was somewhat redundant—there is no way to stop and refuel between the mainland and Hawaii!)

The "Dole Race" was won by Art Goebel from Los Angeles, the same man who had flown my grandmother over Los Angeles in 1925. Years later, Colonel Goebel was a passenger on a B314 Clipper I was working on, flying the same route from Honolulu to San Francisco. On these long flights, it broke the monotony to invite distinguished passengers to the flight deck. They invariably asked routine questions like, "How are we doing?" and "Where are we now?" and we'd impress them with out knowledge. But when Goebel came up to talk to us, we were the ones who were impressed. He told us about his first crossing and how it compared to the comfort and speed of the modern Clipper. Finally, I asked him what had impressed him most about his original flight. He shook his head and smiled. "I guess," he said, "it was when the sun came up for the second time."

Keeping Fit

During layovers in Honolulu, Pan Am arranged for crews to use the facilities of the Outrigger Canoe Club on Waikiki beach. It wasn't entirely kindness. They figured that volley ball and swimming would keep us fit, and a good suntan could be a life saver if we had to spend several days in a life raft.

The Civil Aeronautics Administration (CAA) required all air crews to carry current health certificates. These had to be renewed periodically by a list of specified Flight Surgeon examiners. However, there was a medical examiner in San Jose who had acquired a reputation for never failing any flight crew man on a periodic physical check. Of course he attracted all the "marginal cases," as well as others in perfect health who didn't want to take any chances. This finagling subverted the intentions of the CAA, but I don't know of any instances where someone failed to perform his duties aloft.

Excessive time off was either a blessing or a curse. In any group of men with ten days free between trips, you'd expect

NATS Packet

7241

"DIAMOND HEAD......AND TURN LEFT"

NAVAL AIR TRANSPORT SERVICE COMMAND, PACIFIC WING   SEPTEMBER 1945

CORONADO PB2Y3

The Coronado with Diamond Head in the background.

that some might do something besides maintaining their health. A few aged rather quickly, with symptoms indicating they might not pass their next physical examinations. However, a few used their time profitably to invest in real estate or business and became quite prosperous.

A Grandiose Prediction

One of the assistant managers at the Moana Hotel was a man named Lyle Guslander. In the blackout room of the hotel one evening, he made an improbable prophecy: that, with air transportation available after the war, the other islands in the Hawaiian group could become popular tourist spots if hotel facilities were available. Guslander saw that they were, masterminding the creation of a number of famous resort hotels.

CHAPTER 6

Island Hopping to Australia

> *We cross the Pacific by "island hopping," stopping every thousand miles or so at one of the countless islands or coral atolls, some of them incorrectly charted.*

I was simply unprepared for the great beauty of the Pacific coral atolls as seen from the air. After the constant drabness of the ocean, suddenly we would come upon intense and varying shades of blue and green, the shallow waters of the coral atolls. Palmyra, Canton, and Funafuti were breathtaking— pure white coral growths that built themselves upward from the rims of submerged volcanoes.

Coral reefs and islands are actually living (and dying) organisms that grow upward and outward in the warmer surface waters. When they start at the circular mouth of an undersea volcano, they can form a doughnut-shaped island with a lagoon at its center which is ideal for seaplane operations. Reefs are delicate. Changes in climate, water, or usage can adversely affect growth. In time, they can even disappear.

A number of our first Pacific stops were within these rings of coral, only a few feet above sea level, and often very narrow. In due course, the military built landing strips on the compacted coral and their natural charm faded.

When Pan American was planning commercial Pacific Ocean service before the war, landing and refueling rights were paramount. Small islands which previously had little economic value suddenly became very important.

The island stops between Hawaii and Brisbane, Australia included Palmyra, Canton Island, Fiji, and New Caledonia, with occasional side trips to Espiritu Santo and Wallis Island.

PALMYRA

Palmyra was the first island stop south of Honolulu. It had been a typical uninhabited coral atoll. Rumor had it that an American from Santa Barbara, California owned it. He'd planted it with coconut trees, planning to return in a few years to harvest them. Before the Navy occupied Palmyra, there were no structures. The Navy put up some huts, including a small frame movie house.

The Marines in charge of the island were very serious about security, including blackouts. If anyone lit a cigarette, a Marine would threaten to shoot it out of his hand. (One of the ironies of wartime was that such blackouts were vigorously enforced everywhere except in the actual fighting area.)

After longer-range Coronado patrol aircraft were available, Palmyra was no longer needed. I hope it has regained its former pristine beauty.

CANTON ISLAND

Canton Island, a thousand miles south of Palmyra, was a perfect choice for refueling on the projected Honolulu, New Zealand run. It is a typical coral reef about twenty miles around. The coral doughnut around the lagoon varies in width from several hundred yards to considerably less, with the highest point only a few feet above sea level. In a typhoon, the entire surface could be covered by waves.

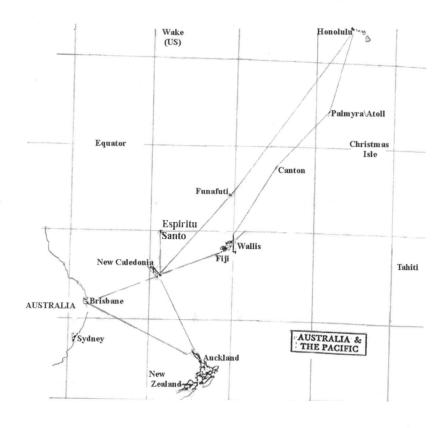

When we first arrived, the lagoon looked like a huge aquarium, alive with tropical fish. One variety swam so fast that, when it broke the surface, it seemed to fly for many yards, sometimes landing with a thud in small boats. The beaches were infested with hermit crabs. These small creatures have no natural shell for protection, so baby crabs find an unoccupied seashell and live in it. As they grow, they keep switching to ever-larger shells. They resist leaving their shelters, even to search for food, and drag their temporary housing along. A number of personnel kept these harmless crabs as pets. Life could be lonely between flights.

Only rubber-soled shoes could survive the sharp white coral. There was no fresh water, and we had to distill potable drinking water from sea water. To provide hot showers, brackish water was put through a network of large sun-exposed pipes—a crude solar heating system that worked beautifully.

The British, who were interested in Pacific air routes, also had eyes on Canton. To establish their ownership of the island, they built a small wooden post office there, staffed by a career civil servant. The U.S. disputed this sovereignty claim by sending a Naval submarine chaser to the area. The Royal Australian Navy then sent a cruiser to insure the safety of the lone "postmaster." Finally, it was agreed that the island would be occupied jointly. A British supply ship would drop off supplies to this sole Canton occupant once a year.

When Pan American established their first base on Canton in the late 1930s, they built a small wood-frame hotel with a tennis court for overnighting their Clipper passengers. By the time the Navy took over, Pan Am had installed docks and maintenance facilities including water distillation equipment. These amenities must have been a welcome supplement for the lone British occupant who still maintained his own compound. He was always welcome at the movies shown in the hotel lobby.

Canton, like so many of the coral atolls, had been uninhabited. When the military set up a station there, Gilbert Islanders were brought in to perform housekeeping duties. These pleasant young islanders were a joy to watch as they made life easier for us on our stopovers.

In this mini-paradise, there was one constant reminder of the war. The President Taylor, a cruise ship taken over by the Navy from the Dollar Line, had gone aground on a break in the reef. Now it clung there, a battered ghost ship going nowhere.

Moonlit Serenades

Between Hawaii and Australia, we always took off at sunrise, arriving about noon at our destination a thousand miles away. Between San Francisco and Hawaii, we always took off at sunset, flew all night, and reached our destination at dawn.

Seaplanes could not land in a lagoon at night, so all island-hopping flights took off soon after sunrise. They could reach their next island destination, landing in early afternoon. There wasn't enough time left to reach another island before sunset.

This made for one flight a day, leaving the afternoon and evening for relaxation. The Gilbert Islanders at Canton Island often would entertain the Pan American compound on the tennis court with native singing and dancing. Their instruments were metal containers and boxes which they beat with sticks. Their singing seemed to show the influence of missionaries several generations back. Often these Gilbert Islanders, far from their native land, would shed tears of homesickness.

Moonlit Bombing

Boom!

"What was that?" Now wide awake, we listened for more. Only a droning sound.

"Either we're being attacked or we're shooting at something!"

"Hey, we better go to the lobby and find out."

"No, get to the trenches. Grab a helmet and get yourself a gas mask *now*!"

On a moonlit night while everyone was asleep, the Japanese had dropped a single bomb on Canton Island. It missed the main 10-room frame "hotel" where we were sleeping by about a hundred yards.

Everyone rushed to the hotel lobby where we were hastily issued World War I helmets and gas masks. Then we crouched in slit trenches for an hour or so. Finally, we realized that the bomb was probably from a lone Japanese patrol aircraft that spotted the glow of the coral island in the moonlight. "Good thing we were asleep and couldn't see it coming," someone said. That explosion recreates itself time and again in my dreams, but I experience no fear because I know the outcome.

A popular radio program for South Pacific service personnel was an anti-American propaganda show presided over by Tokyo Rose. She was supposed to destroy American morale, but we thought she was pretty funny. Miss Rose played current American dance music, read the latest "news," and told us how futile our efforts were. She reported the bombing of Canton Island as a great Japanese victory. We all got a good laugh out of that.

Sad Mistakes

Whales, observed from 4,000 to 6,000 feet above the sea, can look like a submarine. I wonder how many whales were bombed during WWII.

He Sells Seashells

I was resting in the crew room one afternoon when a teenaged Gilbert Islander approached me with a sock full of sea shells he was trying to sell. They were beautiful, but too heavy for my limited baggage. (And I'd discovered that recently-harvested shells could be the source of unpleasant odors.) I didn't want to hurt his feelings, so I praised the shells but said I had no money with me. Several minutes later, he returned. He was sorry I had no money, he said, and wanted to make me a gift of the shells. I was really touched. Here was someone with absolutely nothing by our standards, and he was trying to cheer me up. Actually, he figured he was much richer than I was. He had the sea, the coral beach, fish, and friends. I was a poor traveler and didn't even have sea shells. I hope he was not offended because I could not accept his gift.

God Bless the National Geographic

Before the war, there had been no extensive need for information about these islands, so precise locations and channel depths were often wrong. We had to correct island locations on our charts by referring to the more accurate publications of the Hydrographic Office in Washington, D.C. National Geographic maps were also very helpful for visual navigation when tiny reefs could be spotted during daylight to get a positive fix of our position. I have often wondered if The National Geographic Society knew how much they contributed to our safety in those early days.

Navigators noticed that they were about fifteen miles in error when they arrived at Canton island. By checking published coordinates with the Hydrographic Office manual, it was discovered that the island was misplotted on our own plotting charts. Thanks again to National Geographic for confirmation.

After the war, when longer-range aircraft were available, there was no more need for these island facilities as staging and refueling bases. Canton, I am sure, has reverted back to nature. Is the British representative still there?

FUNAFUTI

In 1944, after Pan Am began operating with Coronado 4-engine patrol bombers, we began stopping at Funafuti atoll for overnighting and refueling. The island had a small native population and a lagoon so large that its waters at times were far from calm.

There were no stores and no way to obtain anything from the outside world except from those transiting the Island. For trading purposes, I bought five yards of brightly printed cotton fabric at Woolworth's in Honolulu, enough for one of the traditional Polynesian lava-lavas, a sort of wrap-skirts worn by the men on Funifuti. My colorful cloth was quickly traded for a magnificent mat, and each of us were sure that we had gotten the best of the bargain.

One day I came upon an island boy, about sixteen, sitting under a bush reading Shakespeare. All this on an island only several hundred yards wide and a few miles around. Obviously a missionary teacher's influence had left an impression.

On another occasion, a native boy named Pencil was caught in some minor crime. His sentence was ostracism. Everyone ignored him for the prescribed period of time. Clearly, this was a punishment far more terrible than prison or torture.

Funafuti is sadly remembered for the loss of one of our planes and crew. The Coronado hit the mast of a ship while taking off. Everyone aboard was killed except the steward.

FIJI

The next destination, another thousand miles south of Canton, was Fiji. Although there are hundreds of tiny islands in the Fiji group, its main land mass consists of two islands, Vanua Levu and Viti Levu. Our PBM landed in the bay east of Suva, the colonial capital on Viti Levu. There was always a pungent, smoky, spicy odor saturating the air.

The New Zealand military was in Fiji to mobilize for the expected Japanese attack. Their three British-built Short Singapore III biplane flying boats were permanently anchored near us in the harbor. They looked like World War I vintage, but had been actually been built in 1934. When World War II started, these three old seaplanes were the only air defense for the Fiji Islands.

The Fijians were traditional warriors with magnificent physiques. We found them to be enormously friendly. No Fijian would pass us by without expressing the traditional greeting of "Bula Bula" with the biggest smile imaginable. Supposedly they were once cannibals, but the missionaries had had a profound

*My official Pan
Am file photo.
(We all had to have
such portraits taken,
just in case we became
"newsworthy.")*

influence. One afternoon I was walking in the dock area of Suva and heard a five-year-old Fijian boy singing to himself. It took me a minute or so to recognize the song: "You Are My Sunshine."

Men, women, and children all wore their hair in a beautifully shaped helmet the size of a basketball, probably as insulation against the weather. They were barefoot, and the men wore lava-lavas. It was obvious that here was a race of people completely at ease with their environment. If anyone accumulated more than he needed, less well-off relatives would move in and consume the excess. They were great at singing and dancing, and always seemed to enjoy themselves. A tropical climate tempered by proximity to the ocean, plentiful seafood, coconuts, pineapples, and yams, made for a stress-free, happy people.

Queen Victoria's Bust

Legend has it that, during Queen Victoria's time, an American sailing ship was plundered in Suva Harbor. The American owners demanded a sum equivalent to $50,000 for the loss. The Fijian chief probably knew little about money. Not having any, he offered to become part of the United States to settle the grievance. His offer was rejected. Not long after that, a British naval vessel anchored in Suva Harbor. When the native chieftain was invited aboard, he noticed a bronze bust of Queen Victoria in the Ward Room. Quite logically, he assumed that these people were governed by a dark-skinned sovereign. He decided that if he allied with the British, they might pay off the Americans. They did, and the Fiji Islands became a colony of Great Britain in 1874.

Paradise Compromised

The British soon realized the vast potential riches they had acquired. Pineapples and sugar cane could be grown, gold was there to be mined, and products from the sea were at hand. However, the Fijians enjoyed their way of life and had no intention of becoming laborers in the fields and mines. The Colonial Government had to import indentured workers from India where there was an abundant supply of poor people eager for a chance at a better life.

It was agreed that the native Fijian should be protected, so laws were enacted which prevented a Fijian from selling his lands. The East Indians became share croppers, and the Fijian continued his traditional way of life, but now with an income from the labor of the Indians. The Indians worked their tails off and shared with the Fijians. Except for some British trading companies in the 1940's, all shop keepers were Indian. The police were Fijian, proud, imposing figures in dark blue lava-lavas with their hair in traditional fullness. The three cultures—European, Fijian, and East Indian—appeared to be compatible.

Fijian Hospitality

We billeted in the Grand Pacific Hotel on the beach in Suva, across the road from the government buildings. The hotel was as British-Colonial as you could imagine. Meals were served in a formal dining room, and tea in the lobby sitting room. East Indians did the housekeeping, but some Fijians acted as waiters. I remember being served soup with a large Fijian thumb solidly immersed in the hot liquid.

We slept on the second floor balcony with mosquito nets over bed frames. Our luggage sat beside the beds, and, to the best of my knowledge, no one ever lost anything from theft. However, there must have been some sort of lawbreaking on the islands because mimeographed notices were posted throughout Suva, listing the fines for various offenses.

One evening in 1943, a Pan American commissary man took several of us into the native area. At first, the Fijians were quite shy, but soon they invited us into their homes. Their houses were constructed with beautifully woven mats, pretty much open on the sides for air circulation. After introductions and some conversation, they started to sing. It was a strange sound, yet familiar. They were singing English hymns that had been taught their ancestors by missionaries many years before, but their harmonies were very different. It was really enjoyable.

I was relieved that we did not progress to the stage of hospitality that included *kava*. This native drink was prepared by the women who chewed the ingredients and spit them into a common vessel for fermentation. The rumor was that it was paralyzing.

Fijian Warriors

The New Zealand military, for obvious reasons, was determined to prevent a Japanese foothold on these islands. The Fijian was a born warrior and took well to military training for their own defense. They didn't march like we did. Instead, they jogged with a great deal of endurance. I understand that they terrified the Japanese in New Guinea.

In one training exercise, a group of Fijian trainees was told to try to infiltrate one of our encampments at night. To avoid accidental bloodshed, the invaders weren't issued weapons. Instead, they had sticks of chalk and told to put a cross on anyone they contacted and presumably could destroy. That night, everything was absolutely quiet. No invasive action. Nothing. Obviously our sentries and precautions had worked.

But as soon as the sun came up, we could see that the entire encampment had been chalk-marked.

*   *   *

After the war, a large commercial air field was built many miles to the north of Suva, at Nadi. Ultimately, the Fiji Islands became independent in 1970. The East Indians became a majority, with intermarriage complicating who exactly was whom. Perhaps Suva still retains its charm. I hope the natives got a square deal.

ESPIRITU SANTO

I had never seen such heavy rain as I did on Espiritu Santo—absolute deluges. The island was located about 600 miles northwest of Fiji, in the New Hebrides group. It wasn't a regular stop in the island-hopping chain, but we often made a side trip there with personnel or supplies.

The first time we landed there in the PBM seaplane, there were no docking or living facilities. We secured the aircraft to a floating buoy, and a Navy launch took us to a yacht formerly owned by the Wrigley family. Perhaps they still owned it, but had made it available for the war effort. The yacht was being used as a temporary billeting facility until the Navy Construction Battalions (called "CBs" or Seabees) could build land facilities. We were instructed to leave any laundry on our bunks for overnight service. The constant contrasts between hardship and luxury were never-ending. (On our subsequent flights to Espiritu Santo, the Navy had set up docking facilities and Quonset huts for operations and billeting.)

I was taking a stroll late one afternoon and wandered a bit too far from the base. Suddenly, about ten natives stepped out of the dense forest on the path in front of me. They had bones through their noses and were carrying spears. Fortunately, they were just going fishing and passed me with a nod. Another time, while exploring the jungle, I found another navigator, John de Kramer, painting a picture of a flower. Constant contrasts!

Michener's Vivid Imagination

A short distance from our operations base at Espiritu Santo was a coconut plantation, run by a bedraggled Frenchman. He lived what looked like a very lonely and primitive existence. His only companion was a nondescript middle-aged Tonkinese woman—whether she was servant or mistress we never knew. His automobile was ancient and dilapidated.

A few years later, when I read James Michener's best-selling book Tales of the South Pacific, I realized that one character was inspired by this same Frenchman. Michener's book was turned into the Broadway musical South Pacific with opera star Ezio Pinza playing the French planter. Subsequently, a Hollywood film was made with handsome Rosanno Brazzi in the role. Michener certainly took a lot of license to make his characters glamorous.

Those Incredible Seabees

I was always impressed by the Navy Seabees. They could make electrical insulators from beer bottles and create culverts under roads by welding oil drums together. Hot water was always available from their makeshift solar heaters. Seabees were recruited from civilian carpenters, plumbers, mechanics, and heavy equipment operators. These fellows could build runways, roads, electrical transmission lines, and set up corrugated steel Quonset huts almost overnight. In some combat areas, the Seabees' bulldozers were used to seal up caves rather than attack the enemy inside.

As soon as an island base was secure, you could be sure that one of the first things the Seabees would do was to construct an Officers Club and one for the Enlisted personnel. These makeshift facilities were very important for morale. After a long flight, a couple of beers would help you relax and sleep better.

The clubs posted rules of a sort. Here's an example:

U.S. NAVAL OFFICER'S CLUB

ESPIRITU SANTO, NEW HEBRIDES

SLOT MACHINES:
Experience has proven that drinks poured into the
slot machines will not help them to pay off. Please
do not shake or tamper with the machines. Use only
in the prescribed manner. If slot machines are not
working satisfactorily, please see the storekeeper
at the desk.

SINGING:
Singing in the proper manner is approved; however
selections are to be in keeping with normal Officer
conduct.

CONDUCT:
Beware of alcoholic promotions, especially those
which elevate you momentarily above the SOAP. For
such promotions are less than temporary, and what a
demotion! There are adequate approved toilet
facilities on the Club premises, so be smart--use no
makeshift. Remember always that you are a gentleman
and an officer. If you must relax completely, do so
gracefully.

NEW CALEDONIA

There was a strict blackout on all island bases to the
north including Hawaii, but here on New Caledonia there was no
attempt to keep the lights off. Activity was on a twenty-four
hour schedule, with brilliant floodlights dominating the night.

New Caledonia was a French possession. It's main
reason for being was its vast deposits of nickel ore. A large
smelter in Nouméa employed hundreds of indentured Tonkinese
workers, many of them women, and its huge smokestack
belched clouds of eye-irritating smoke.

The Pan American Naval Air Transport facility in New Caledonia was located on the island of Ile Nou at its capital, the typical French colonial port city of Nouméa. The Navy was using it's magnificent and very large bay as a repair facility and seaplane base. Nouméa's harbor was protected by a series of nets which, when deployed, would theoretically bar enemy submarines from the bay, even the midget ones that carried a two man crew. One or two of these small submarines may have gotten into the bay. In any case, they were blamed when the ammunition depot at the smelter blew up, causing about 250 casualties.

Ile Nou was a very important supply point and headquarters for the Construction Battalion (Seabees). Admiral William Frederick "Bull" Halsey kept his PB2Y3 Coronado anchored next to our facility. His headquarters were at Nouméa, a mile away.

New Caledonia isn't a coral island. It is about 220 miles long and forty miles wide, with a continuous range of mountains along much of its center. The climate is similar to that of Hawaii because it's as far south of the equator as Hawaii is north. Beautiful flame trees contrast with its grim history as a Devil's Island. The old prison administration buildings were renovated and used for operations and billeting. There were still ominous and decaying cells dug back into the mountain.

With its great beauty, Ile Nou could have become a great tourist attraction. However, the increased long-range capacity of aircraft eventually led to its being by-passed.

Jello, Spam, and Bing Crosby

On my first PBM trip to New Caldonia, I found that the only food available at the Navy mess was Spam and Jello. Breakfast, lunch, and dinner. It had been like this for some time. The next day, our purser arranged for us to go aboard our own anchored seaplane where he had saved some state-side steaks. It was a bit incongruous to be served thick, juicy T-bones by a well-known San Francisco chef in the middle of Nouméa Bay, when the military regulars weren't even getting creamed chip beef. But we managed to suppress our guilt and greatly appreciated his efforts.

The intense displacement of wartime produced frequent incongruities. During the tropical Christmas season, state-side holiday music and snow-filled movies were in abundance. Perhaps as a cynical antidote to homesickness, one of crooner Bing Crosby's most popular songs was altered to "I'm Dreaming of a White Mistress." That always got a chuckle.

On another occasion, our captain, Dick Scott, felt that some minor repairs on a PBM seaplane justified a test flight in Nouméa harbor. We invited two natives to go along on what must certainly have been their first flight. It was a wondrous day for them, and hopefully gave them some status among their peers.

Suspected of Spying

I got in the habit of taking a book or current reading material with me to make layover times more productive. The Russians had become our allies in the war. Anticipating that Pan American might expand into Eastern Europe after the war, a group of us decided that a knowledge of Russian might prove useful. Several of us started taking lessons from a tutor at the University of California. This would be a good use of those long hours in some hotel or barracks while waiting for the next airplane. All was going well until I arrived at Customs in Honolulu on a flight from Australia. The Customs Inspector couldn't read my class notes and decided they might be coded messages. He confiscated all my study material. I never went back for further lessons.

Security was naturally a primary concern during the war. Midway in the Naval airlift in the south Pacific, all crews were commissioned and inducted into the Naval Reserve. We were ordered not to carry cameras, keep a personal flight log, or talk about our routes. When our plane returned to San Francisco, Pan Am would take away its log. After the war, we were all given copies of our flight times and itineraries. But I suspect that Tokyo Rose knew them all along.

Paradise Lost

World War II seaplanes touched down on remote islands and set up facilities, with devastating effect on native populations and their environment. We ravaged their ecology, sometimes changed their dietary habits, introduced diseases they had never known, and then abandoned them when their tiny atolls were no longer needed.

Small islands like Okinawa and Iwo Jima were worth major bloody battles when they were needed for temporary refueling stops. Now they would not be given a second thought. Long-range aircraft eliminate the need for thousands of ground troops to capture and hold refueling bases. If World War II had been fought twenty-five years earlier or later, they would have been left to their peaceful, pleasant ways of life. Then it took six or seven days to fly from San Francisco to Brisbane, Australia. Today a space capsule does it in almost as many minutes.

Now that jets can cross the Pacific nonstop, many of these once important fueling stops are no longer necessary. This is a godsend for the islands and native peoples who, I hope, have recovered from their brush with "civilization." What wonderful, simple people they were when it came to sailing, fishing, and knowledge of their environment. I felt very inadequate.

BRISBANE

The final leg to Australia took us to Brisbane in Queensland. In those days, it was a very provincial seacoast town, about fifty years behind state-side when it came to amenities.

General Douglas MacArthur had his headquarters in Brisbane. One of our captains, a swashbuckling former all-American football player named Steve Bancroft, did personal shopping for Mrs. MacArthur in San Francisco. This gave her a source for goods unavailable in Australia, and, in exchange, Steve got special billeting in Brisbane.

A Race to the Bathroom

On my first trip to Australia, I joined the other crew members in rushing to the bathroom. We weren't sick. We were just eager to make a scientific observation. We'd been told in school that water spirals down the drain in a counter-clockwise direction in the northern hemisphere, but reverses to clockwise in the southern hemisphere. Guess what? It does.

GI's and Aussie Girls

American GI's were being sent to Australia to protect the country while their men were fighting in north Africa. The young Australians had been gone far too long, and American service men found themselves in great demand with Aussie girls. It had even become customary for local ladies to go to certain hotel lobbies at 5 P.M. and line up to meet Americans.

Australians are great at water sports, and one of their better beaches was near Brisbane. While we were sunning ourselves one afternoon, we watched an extremely tall young pilot checking out the local ladies for a potential dance partner. This guy was really crazy about dancing. Eventually he focused on an attractive girl who was sort of buried in the sand. He turned on his engaging American charm with this Australian lass until she agreed to a date. We were all taken aback when she brushed off the sand and stood up. She was about four-and-a-half feet tall. Barefoot, she looked straight into this guy's belt buckle. I can't recall whether he wiggled out of it some way, or if they did go dancing. I'd love to have seen it.

When the Australian soldiers got back, they rightfully resented our popularity. Americans had more disposable money, which made matters worse. The returning Aussies were seasoned fighting men, very, very capable. If several of them spotted a lone American, they sometimes gave him a good trouncing. We understandably gave them a wide berth if we saw them coming.

CHAPTER 7

Champs Élysées,
South Pacific Style

> *I discover that the wartime south
> Pacific is the crossroads of the world
> for encountering old friends and
> making new ones.*

They say that if you sit at a sidewalk café on the Boulevard Champs Élysées in Paris for ten minutes, someone you know will walk by. I had the same experience in the south Pacific during the war.

We covered a lot of territory flying around the South Pacific, while many wartime personnel were more or less stationary. This meant that I was always coming across people I knew. I ran into my UCLA Spanish teacher and my high school mechanical drawing teacher, Major McGee, both in Hawaii, and a college friend in New Zealand. Military "courtesies" were greater than I expected, which allowed me to track down and visit with old friends.

John Patton

On one of my first trips to Nouméa, it looked like the whole Pacific Fleet was in the harbor. I had heard that a high school classmate, Navy Lt. John Patton, was aboard the carrier Enterprise, a ship that had recently taken some serious hits in battle. I got permission for a sailor to take me to the carrier in a small launch, walked up to the flight deck, and asked to see my friend. After several minutes, I was escorted to his quarters where he was recovering from his injuries. The launch stood by to take me back to Isle Nou, and within a few days I visited John's mother in Turlock with the good news that he was alive and well. John later became an Air Force Major General.

John "Red" Hansen

During a layover at Brisbane in 1943, I rode a narrow-gauge railroad train thirty or forty miles inland to visit a school friend who was stationed at an American training base. John S. "Red" Hansen had been a high school football player and a member of the University of California Varsity rowing crew. He was more of an energizer than a scholar, and is remembered for his ability to create team enthusiasm. He had paid his modest college expenses by waiting on tables.

Although John was not adept at marching, he was now training officer candidates to do just that. Again, his booming voice, honesty, and energy were paying off. We had lunch that day with his Commanding Officer. This man had encouraged John to correspond with his daughter. Later, the two married and settled in the Portland, Oregon area where he parlayed a career in radio into management of a television station.

Again, as soon as I was back in California, I visited his mother in Turlock to reassure of his well-being. "Red" Hansen proved that a burning enthusiasm was probably as important as scholarship in both war and business.

A New Cast of Characters

The people we work with in our youth, especially in wartime, always seem bigger than life. Yet, I'm sure that many of my colleagues during those years are truly unforgettable people.

Steve Bancroft

Some pilots excel in the engineering and more scholarly aspects of the profession, but fly "by the book," somewhat mechanically. Others get into an airplane and seem to wear it like a coat. They are natural-born pilots, likely to be quite athletic and not particularly "ground school" oriented.

Captain Steve Bancroft was one of the latter. He was a football celebrity, supremely confident and charming, who enjoyed every minute of living. He was well known for his numerous practical jokes. U.S. Customs in Honolulu had been growing more and more irritated by his cavalier attitude and were determined to catch him at some transgression.

Steve decided to have some fun with them. He arrived from the south Pacific and, following procedure, put his bag on the customs table for inspection. Tucked under his arm was a neatly wrapped shoe box, which he hung on to. The Customs Officer inspected Steve's bag, then said, "And now, Captain, what is in that box under your arm?"

Bancroft replied, "Oh, just some horse s---."

"Well we will just take a look at that horse s--- if you don't mind."

The box was inspected. Thoroughly. It contained just what Steve had said. The Customs Inspector had to rewrap the package and return it to the chuckling pilot.

Bob MacGregor

Bob, a Pan American traffic manager, had been stationed in Manila before the war. He had been called back to the U.S. for a meeting when the Japanese invaded the Philippines, but his wife was captured. She was held in the Santo Tomas prison camp throughout the war. (Happily, she survived and they were reunited after the war.)

MacGregor was quickly assigned to traffic and cargo duties in Brisbane, Australia. He too had dreams about post-war Hawaii. During one of those long evenings in Brisbane, he told me he intended to get a piece of the action.

A lot of guys talked about their hopes for the future, but Bob did more than talk. When peace came, he established Trade Wind Tours and became a leading businessman in Honolulu. Later, he sold out for a sum he couldn't resist.

Thirty years later, I was on the inaugural Boeing 747 Braniff Airways flight from Dallas to Honolulu. Many celebrities from Dallas and Honolulu were honored guests. One young lady was wearing a Trade Wind Tours badge, and I couldn't resist telling her about the dream of Bob MacGregor. She confirmed my remembrance. "Yes," she said, "Bob was my father."

Ferd Kline

A German Olympic ski jumper in 1936, Ferd defected to the U.S. to become a flight engineer. He later operated a sporting goods store in the Fisherman's Wharf area of San Francisco for many years.

Stanley Smith

Our navigation instructor, Stanley Smith, made the longest flight between San Francisco and Honolulu; more than 27 hours.

Captain Bob Ford

Bob's unscheduled round-the-world flight was front page news. He was in the New Caledonia area, flying a Boeing 314, when he learned of the Pearl Harbor attack. He was scheduled to return to home base via Hawaii, but that was now definitely inadvisable. He decided to fly in the opposite direction, west to

New York via Australia, Africa, and South America! His heroic flight in complete radio silence and secrecy was the first time a commercial plane had flown around the world. Ford, a real gentleman, was always fun to be with. He had a disagreement with Pan American in the jet transition period and left the company, ending his illustrious career with the non-skeds. I flew with him a lot.

Almost Alaska

Alaska in the 1930s had held a great interest for me. Actually I had seriously planned to study geology and mining at the University in Fairbanks, so I had accumulated a number of books on what was then a territory. One of the books was about well-known "bush pilots" in the Arctic areas.

In 1944, Pan American bought up a small Alaskan air line to acquire Alaskan routes. The Pan American Pacific Division became the Pacific Alaska Division. To accomplish the merger, the pilot seniority lists had to be integrated. The Alaskan pilots were small-plane qualified and very skilled at what they had been doing. But suddenly their seniority earned them assignments on large planes, providing they could be trained quickly. Naturally, their backgrounds did not include oceanic navigation. I went with one of these pilots on his first flight, and was pleased that he requested me again and again.

Pilot Overboard!

Usually everyone was serious about these long ocean trips, but sometimes boredom produced horseplay. Once, when a navigator had painstakingly taken and plotted a three-star fix and someone cut it out of the chart with a razor blade, so that the process had to be repeated. Then a blob of chewing gum was affixed to the chart right on the course to be flown. The navigator said nothing, but navigated to the "gum mountain," then gave compass headings to enable the plane to go around the blob. And on several occasions when we had brand new stewards on board, the captain would switch on the automatic pilot and open the windows. Everyone would conceal themselves in the hollow wings, on the top deck, or in the sleeping room in the back. Then the captain would ring for the steward and quickly hide. The guy would come to the cockpit and find that no one was there. Terrified, he would race back and tell the purser that no one was flying the plane. The purser would come running, only to find everyone back in their usual positions, acting like nothing had happened.

A Pyramid in the Cockpit

I don't know the derivation of the "Short Snorter" Club. It was a ridiculous penny-ante pyramid scheme, designed to be an ice breaker in strange ports. To be a member, you paid a dollar to everyone present—usually two to six guys. Then you produced another dollar and they signed it as your "recruiters." As a new "member" you were now authorized to recruit others and receive their dollar bills.

Long flights are boring. The number of military passengers was usually limited, so it was not unusual for the stewards to offer a cockpit tour to Short Snorter members. If someone was not yet a duly authorized member, the steward, of course, could arrange it. Another facet of the game was to challenge someone to produce their Short Snorter bill within a short time. If they couldn't, they had to purchase a round of drinks. You tried to make sure that you took your bill with you everywhere, even to the beach.

When a bill was covered with signatures, you'd start another, taping it to the first one. Eventually I had one ten feet long, rather a large bulge for my wallet. "Would you please sign my Short Snorter?" was a great way to get celebrity autographs. This pyramid scheme produced thousands of World War II souvenirs.

CHAPTER 8

Trouble at Sea

> *Three tales of near-disaster*

In my 15,000 hours in the air, I've been extraordinarily fortunate. However, the chance for something to go wrong was always there.

Eddie Rickenbacker Lost at Sea

It was front page news nationwide when Captain Eddie Rickenbacker, World War I flying ace, went down in the Pacific, possibly due to a navigational error. (The word among navigators was that an octant had been damaged.) Rickenbacker, founder and CEO of Eastern Airlines, was now a civilian, making an official survey of south Pacific aircraft facilities. By 1943, a few coral islands had landing strips.

Fortunately, Rickenbacker and the military crew took to their life rafts, and were rescued after nearly a month at sea. Rickenbacker attributed the rescue to "divine intervention." So, the navigator's gear on life rafts was replaced by prayer books.

I was navigator on the Boeing 314 that brought Captain Cherry and his crew back to San Francisco. Rickenbacker had previously flown home in glory. There was considerable resentment among the young military crew over Rickenbacker's domination of the incident in the press: "We stayed alive on that raft just to watch that S.O.B. get all the credit!"

A Near Miss

A Navy Admiral needed to get to Auckland, New Zealand. When we got the assignment, we were told it would be the first Naval Air Transport flight over that route. (I recall that the Admiral was very upset because the Japanese had been dropping pornographic pictures to the troops, telling them that their girl friends back home were misbehaving.)

As we approached Auckland's Mechanics Bay, the weather was overcast. The pilot decided to follow a radio-range leg to ensure a safe letdown through the clouds below. Correct procedure is to request clearance by radio so you know that no one else is doing the same thing. To do otherwise is an invitation to disaster. The young copilot repeatedly reminded the pilot that he did not have clearance. "Shouldn't we let them know?"

This pilot was used to having the oceans and airways to himself. An unfortunate arrogance had developed, an attitude that "No one on the ground is going to tell me what to do." We made it down. But as we docked the plane, we were met by the New Zealand Civil Air Authorities who were very upset over the incident. Another airplane had been flying in the same area. We had endangered both ourselves, the Admiral, and another aircraft.

Our pilot kept insisting, "I didn't have time to follow your procedures." Perhaps he resented the much younger copilot telling him what to do. Over the years, although this pilot climbed the executive ladder, I noticed that his character remained flawed. To put it politely.

Lost in the Fog on Christmas Eve

San Francisco Bay is often covered by dense fog in the morning. It was particularly frustrating to arrive there after an all-night flight and not find a safe place to land. A sea plane, by

Boeing 314 Clipper in San Francisco Bay.

its nature, can't "land" on land. Alternate water-landing sites included Clear Lake to the north, Tulare Lake to the south, and perhaps the Sacramento River.

On Christmas Eve, 1944, after an all night-flight from Honolulu, our Coronado seaplane arrived over the Bay early in the morning and could not land due to dense fog. To circle the area waiting for a break could consume the fuel we needed to go on to an alternate landing site. Then the pilot spotted a slight break in the fog in the Benicia area, east of the Carquinez Bridge. The plane landed, and the sea anchor was lowered. Several other seaplanes, also inbound from Hawaii, heard we had landed and followed us down.

By noon, the fog was still thick, so the idea of flying down to a new base at South San Francisco wasn't practical. A few hours later, visibility improved, and it looked like we could taxi the plane forty miles to home base. As we were passing under the Carquinez Bridge, we noticed that the other seaplanes were following our lead. Then visibility started deteriorating fast. We had become, for all practical purposes, a boat competing with any marine traffic in the area. In the zero visibility, our plane drifted toward shore, damaging one of the pontoons.

Our cargo consisted of mail bags. The crew clambered out and fastened a number of them to the top of the wing opposite the damaged pontoon to counterbalance and help the remaining pontoon stay in the water.

Suddenly, a Coast Guard cutter appeared through the fog. They put a cable around the bow post of the plane and tried to tow us to San Francisco. However, the bow post had not been designed for towing. The plane zigged and zagged behind the cutter. Then it crashed into it, opening a hole in the front of the plane's hull. We sealed the bulkhead doors to prevent sinking, and I spent the next few hours holding a series of rolled-up blankets against the hole. Another problem was that the Coast Guardsmen forgot that we had wings which could hit pilings as they towed us.

After more several delays in the heavy fog, the lights of San Francisco became visible late on Christmas Eve. The Coast Guard bid us farewell, and we started up our engines to taxi home.

It was a weird looking airplane that was delivered to it's home base by an exhausted crew on Christmas morning. The forward section of the plane was not buoyant due to the hole up front, so the tail section rode unusually high. It seemed strange that, after operating all over the South Pacific in wartime conditions, my only real emergency was down-at-sea in San Francisco Bay, almost within sight of where we got our emergency training. On sea planes, the captain was often referred to as "the skipper," a nautical term. It seemed even more appropriate after this incident.

We had not eaten or slept for two days, and knew that the exposure would take its toll later. I arrived at my domicile in Berkeley, and immediately piled into bed. Within an hour, the telephone rang. It was my mother in Turlock saying, "Aren't you coming home for Christmas?" I didn't disappoint her. I got up and drove the hundred miles to a wonderful Christmas dinner. Several days later, I reported to the Company Flight Surgeon and found the rest of the crew there with the same problem: exhaustion.

CHAPTER 9

End of the Seaplane Era

After World War II, land-based planes replace seaplanes, as Pan Am initiates an Africa-Orient Division to bring U.S. troops home from India, Europe, and Africa.

By 1945, the era of the pioneering Clipper aircraft was history. The ocean was no longer ours exclusively. A huge flying boat, the Mars was now in operation between Hawaii and the west coast, and the Navy had developed its own squadrons of both land-based and seaplanes.

Three Coast Guard cutters were permanently stationed between Hawaii and the west coast, so that aircraft were never more than 250 miles from a rescue ship. Radio bearings from these Coast Guard ships were usually available for navigation as planes came abeam. A new electronic positioning system called LORAN was now being used.

Airfields had been constructed on most of the islands, as well as throughout the continental United States. Our lumbering seaplanes, which could take off and land only in the daytime, were being outperformed by both Navy and Army land planes, which could cut delivery times in half by operating around the clock.

The End of the War

V-E Day—Victory in Europe Day—was May 7, 1945. The war in the Pacific was rapidly moving toward Japan, and flying the Pacific had become routine. By June of 1945, Pan American's Africa-Orient division, operating as the Air Transport Command, was expanding to bring the troops home from India, Africa, and Europe. It needed a few more navigators, and I was chosen to be one of them.

In August, 1945, while I was driving to my new assignment in New York, atomic bombs were dropped on Japan, effectively ending the Pacific war. When I got to New York, my first subway ride was from Queens into Manhattan where an enormous celebration was going on in Times Square. I had arrived on the evening of V-J Day marking the victory over Japan. I came up out of the subway to find myself in the middle of tens of thousands of ecstatic people. Everyone was dancing and leaping about, laughing, crying, hugging and kissing each other. After nearly seven exhausting years and twenty million dead, the slaughter was ended. The war was over.

Across the Atlantic

My new assignment, the Africa-Orient Division operated out of the Marine Terminal at La Guardia Airport. Our mission was to provide airlift in and out of Bermuda, Newfoundland, the Azores, London, Paris, Frankfurt, Rome, Prague, Athens, Casablanca, Cairo, Baghdad, Karachi, New Delhi, and Calcutta. Reliable C-54 (DC-4) aircraft were used.

Bigger Planes, Smaller Cockpits

The flight decks on seaplanes had been all spaciousness and comfort, with large navigation tables. Compared to them, the DC-4 cockpits were cramped. There was little room to stand up and walk around. The navigation table was a makeshift piece of plywood mounted by the cockpit-access door. There was a tiny stool to sit on that doubled as a foot stool for star viewing. However, averaging octants (dubbed "mixmasters") could be hung in a plastic astrodome. Finally we could operate without having to open hatches and expose ourselves to cold air!

There was a certain irony. As each generation of aircraft grew larger and faster, the crew got less and less space. Navigation facilities got smaller and smaller until, finally, the navigation station was eliminated entirely on the Boeing 747 jumbo jets.

First Time on Wheels

This was my first experience with a plane that could not land on water. In the winter, we would carry Arctic clothing for flights going through Newfoundland and Europe—a lot of bulk—plus tropical gear for east of Cairo. Now we were allowed to carry cameras. Crews were scheduled for a flight every five or six days, so we had layovers long enough to enjoy and explore all those delights that the National Geographic had been promising.

It seemed eerie to be in Germany and Italy so soon after the war. Rubble from bombing was still along the streets of Frankfurt. The I. G. Farben Building had not been bombed because the military hoped to use it as headquarters after the fighting stopped. In London, our hotel still had broken windows.

Inflation and scarcity were particularly noticeable in Italy and Greece. Whenever possible, we tried to stay in military BOQs (Bachelor Officers' Quarters)and eat at Officer's Clubs. Our off-duty billeting varied considerably. In larger cities, it was usually a fine hotel like the Crillon in Paris. On military bases, we stayed in BOQs. At a desert station, it would be a fairly rundown facility.

When we made sixty-day trips to India, Pan American gave us a $500 cash advance. Usually there was one flight a week, which meant a seven-day layover for every plane change.

*In my Africa-Orient
Division uniform for
Pan Am, 1946.*

This was a lot of cash to be carrying around in areas where people had been so hard hit by hostilities. Our military-style uniforms afforded some safety. (The major players in the military wore golden designs on the visors of their caps, known as scrambled eggs. The rest of us had no scrambled eggs.) I made two of these trips during my ten-month assignment to Africa-Orient.

A Real Maquis

At the Pan American Marine terminal on La Guardia airport, crews in 1945 had to take a refresher course in the use of emergency equipment. This training was conducted by a Frenchman who had worked as a *maquis* during the war, a member of the French underground. His job was to parachute into occupied France with radio equipment on his back, deliver it to the Resistance, then make his way back to England via the "underground" so he could repeat the process. A very brave man.

The Pope and the Mysterious Package

When a bunch of us were visiting the Vatican in Rome, someone suggested that we try to see the Pope. "Yeah, right!" I thought. But to our absolute astonishment, we were told to come on in. "His Holiness will be right with you." Apparently, the

Pope was taking time to wish uniformed military a safe journey home. We also toured the Vatican Museum, a fantastic repository for history. It would have been an unthinkable loss if careless bombing had destroyed the area.

As we crossed the tarmac to board our plane in Rome, a neatly dressed woman in her forties approached and asked if we would hand-carry an envelope for her and drop it in a mailbox in Paris. Crews had strict instructions not to carry packages or letters for anyone under any circumstances. So we told her we couldn't help her. As soon as we landed in Paris, Military Security came aboard, and said "All right, who has the package?" One of the military passengers, a Captain going home for discharge, had it. He was removed from the flight and would be detained for who knew how long.

We never found out what was in the package.

Over France

When you flew from the Azores to Paris in those days, your first sight of France was near Brest, followed soon by the famous isle of Mont-Saint-Michel. I was always surprised how much its pointed silhouette resembled the Eiffel Tower. From the air, it was hard to tell where the rock stopped and the ancient fortress and abbey began. This rocky crag sits at the end of a spit of land, its causeway submerged at high tide so it becomes an island. From there to Paris, we flew over fields and towns scarred by hundreds of bomb craters.

Horse Play

One evening in Paris, we stopped at a small bistro for dinner. Someone in our group had said, "I'm hungry enough to eat a horse." Later we found out that maybe we had. Not because of the war, but because it is a traditional French dish.

At the Louvre, we were surprised how casually the most famous art works were displayed. In a hallway near the entrance, with no fanfare, hung the Mona Lisa. (She is now behind bulletproof glass on the second floor.)

Buttered Up

An upscale photography studio stood near the Madeleine. We decided to have our portraits taken in our uniforms and pick up the prints the next time we returned to Paris. The manager was a lady in her mid-forties, charming and obviously well-educated. She told us about her experiences in the French underground during the German occupation and offered to be our guide to Versailles if we could provide a car, which we did. In return, we invited her to have lunch with us at our hotel, the Crillon on the Place de la Concorde.

Paris was not yet back to its standard fare, and this invitation was special for her. She felt obliged to bring a "gift" for the occasion. When we were seated in the luxurious dining room, she reached into her purse and produced a pound of butter. Later, I sent her four bars of fancy soap as a thank-you for the tour. Almost a year later, I received her response from Africa. She had married a French diplomat and was stationed in the Sudan.

Baritone with Bouillabaisse

We were flying into Paris on Thanksgiving day, 1945, and found that Orly airport was closed down due to weather. We landed instead at Istres le Tube, which had been a training facility for the French Air Force near Marseille in southern France. The weather remained bad for several days, so we had the chance to take a bus into Marseille and have dinner in a beautiful old hotel dining room.

Marseille is a famous seaport with a reputation for roughness, so we felt it best not to walk around the dock area at night. While we were lounging after dinner, we heard an operatic baritone voice coming from the courtyard. We went to a window and discovered that this fine voice was coming from a crippled beggar, singing for coins. His physical condition may have been the result of poverty and lack of medical attention. Though his voice was untrained, he had raw talent. Under different circumstances, he would have a more affluent audience, hopefully indoors.

Surprises in the Azores

The island of Santa Maria in the Azores was used for crew changes on military charters to both Europe and Casablanca in Morocco. The Azores are the Portuguese islands where Columbus stopped on his voyages. The winds there were often excessive, and life was primitive by 1946 standards. Fishing and marginal farming were the norm for these hardworking folk.

I was reminded of the many Portuguese from these bleak islands who had found a better life in the central San Joaquin Valley of California in the early 1900s, becoming substantial contributors to the local culture.

In the Santa Maria airport restaurant, a very young fellow was employed as a waiter. His white apron dragged on the floor and his knowledge of English was marginal, but his enthusiasm for this hard-to-get job was apparent. Several of us ordered scrambled eggs for breakfast. This seemed to create a problem. He said, "Sorry, we are out of scrambled eggs. Do you want them up or over?"

About a mile from the airport in Santa Maria was the village of Vil du Port. Its principal street sloped steeply down to the sea. It seemed to be a fishing town, but the complete lack of commercial activity combined with the persistent harsh winds underlined the fact that these were among the hardiest people anywhere. Except for the electricity, I could imagine Columbus taking on supplies down at the dock. The children were often very attractive, but the standard of living for all was difficult.

At the top of the main street, an enterprising family had opened their modest home to people from the air base who wished to eat a meal off base and on the Azorean economy. There we enjoyed unexpectedly delicious steak, red wine, and wonderful bread in a rustic setting.

Walking Billboards

TWA was just starting service to Cairo. I don't know if it was a clever promotion stunt or they actually had hundreds of employees, but we saw people wearing white TWA overalls all over the city during the few days we were there. Perhaps TWA had hired them as walking advertising. If so, it was very effective.

Something (and Someone) to Remember

I particularly enjoyed a Cairo layover where we did all the expected things like group pictures at the Sphinx and going inside the narrow passages of the pyramid at Giza. It was a walk through time as well as space.

Back in Cairo, several of us needed some help in negotiating the price of a carriage ride to Groppi's, a famous confectioner. An imposing official came over and helped us in the transaction. I complimented him on his resplendent uniform, not realizing that in his culture I was making a request for one of his uniforms. He offered it, and I of course refused.

We had cocktails on the verandah of the Shepheard Hotel in downtown Cairo. I had occasion to visit the men's room, and was greeted by a young, very heavy-set, balding attendant who, in English, asked where we were going. "New York," I replied, and he said, "Have a nice trip." I suspected that he could say these few phrases in many languages, thus obtaining better tips. For some reason, this incident lodged in my memory.

In 1952, the burning of Groppi's and the Shepheard Hotel marked the beginning of the Egyptian revolution against British rule, a prelude to the eventual ousting of King Farouk.

Spam and Dates

From Cairo we flew to Lake Habbaniya, about forty miles from Baghdad in Iraq. It was a British air base and very much a colonial outpost in the desert. We had a twelve-day layover on one trip, with only meals to break the monotony.

Playing tourist with our guide and the Sphinx in Egypt.

Unfortunately, the base was poorly supplied and short of food, probably due to a planned cutback in operations. The only thing we had to eat was Spam in some form or another. All night long we heard wild jackals howling mournfully. If they too were living on Spam, I couldn't blame them.

We engaged a cab to take us for a day-trip to Baghdad so we could see the local bazaars and shrines. I purchased a package of dates from a street vendor and gave him money to mail them to a friend in the U.S. The vendor could have kept the money, certain he'd never see us again, but several months later the dates arrived. The vendor's code of honor would not allow him to do otherwise.

That evening, we visited an Arab "night club" where belly dancers were performing. Our presence made them uncomfortable until we told them we were Hollywood talent scouts. We snapped some pictures and left.

Lofty Impressions

So much history is jammed into such a compact area in the Middle East. Our copilot was an anthropologist, and he was constantly commenting on the fascinating and important sites below us. One trip from Rome to Baghdad took us over the Corinthian Canal in Greece. It was not operating due to wartime damage, but it was still impressive with its sheer vertical walls seemingly carved from living rock. It was obviously a tremendous construction achievement. All that human effort, and then one bomb could destroy it all.

We flew on until we were over Haifa, then veered left to have a five-minute bird's-eye tour of Galilee and the Holy Land. So small, yet it represented the entire world to some of the founders of western civilization. I was excited, thinking, "My gosh, this is it!" I managed to snap two photos of the Sea of Galilee. Then it was behind us as we flew on.

Karachi

In Karachi, India (now Pakistan), we saw many young boys who apparently lived on the streets. Many had a singsong patter which went, "No mama, no papa, no flight pay, no per diem, *kumshaw* [tip] please." As far as they were concerned, anyone in uniform must have flight pay. American service men befriended them and taught them enough English so that they could be useful as protection and in negotiating for small purchases. One boy, called "Chicago Charley," always seemed to be with us when we needed him. With his tips from us and certain kickbacks from vendors, Charley made more than most adults.

An enormous wooden hangar stood at the Karachi airport. It had been built for the projected round-the-world flights of the German dirigible Hindenberg. But the Hindenberg, buoyed by flammable hydrogen instead of helium, exploded and crashed in New Jersey. The flights never took place.

*The giant hangar at Karachi, 1946. It was made before
WWII to house the giant Hindenburg—but sadly never used.*

The area around the airport was very British Colonial. Several of
the crew got sick from the food or water, but the layover was
long enough to recover despite the oppressive heat.

The Top of the World

The first time I flew from Karachi to Calcutta, we
diverted slightly to Agra so we could see the Taj Mahal from the
air. We circled it at very low altitude so we could take
photographs of this exquisite structure.

On my second Calcutta trip, we were scheduled to stop at
New Delhi with a load of China-Burma-India troops returning
home. Because only a light fuel load was required and because
the operation was due to shut down, we rationalized that this
was the time to divert north into Nepal to see Mount Everest. It
was 1946, and Everest had not yet been climbed. The non-
pressurized C-54 airplane got up to 26,000 feet with everyone on
oxygen. Of course we couldn't do more than go abeam of
Everest which is 29,000 feet high, but I did get some pictures of
the "top of the world."

*A guard outside the thatched ammunition depot at
the Barrackpore military airport at Calcutta, 1946.*

End of the Africa-Orient Airlift

The Africa-Orient airlift lasted until mid 1946. Then Pan
American offered some of us overseas ground-management or
dispatch jobs. Few considered these jobs a good career move,
and most turned them down including me. We were mostly
young and just married, so we didn't want to take wives and
children to these remote spots. I was then asked to go to the
Atlantic division which serviced Europe and Casablanca. I
accepted.

The Atlantic Division was a commercial operation using
Lockheed Constellations. The "Connie" was pressurized for
higher altitudes and much faster than anything we'd ever had
before. Higher altitudes put us over clouds, and greater airspeeds
meant fewer hours aloft. Unfortunately, the Connie had a very
tiny navigator station, so I was practically sitting back with the
passengers. However, I had been at Lockheed while this plane
was being developed, so it felt like an old friend.

Faking It for the Tourists

The airport in Monrovia was very near the Firestone rubber plantation. Liquid rubber, tapped from the trees, was processed in a large building which contained dozens of De Laval cream separators, exactly like American dairy farmers used to separate cream. Only the "cream" was high grade latex rubber.

During a layover at Roberts Field, Monrovia in Liberia, a group of us went into the jungle with our cameras and came upon what appeared to be about a dozen deserted huts. After a few minutes, heads began popping out from all around. The local people were wearing GI shorts and white tee shirts.

We asked if we might take their pictures. "Yes," they said, but first they'd have to take most of their clothes off. "The folks back in the States think we are savages, and we don't want to disappoint them." They even simulated a knife fight so we'd have some action photos.

A giant ant hill amid the rubber trees of Monrovia.

"We are not savages." Descendants of American slaves in Liberia.

All these folks had familiar Anglo-American names like "Johnson" and "Smith." They were descendants of freed American slaves who chose to live in Africa after the Civil War. Liberia means free, and Monrovia got its name from President Monroe. The country was founded in 1822 by the United States as a potential home for former slaves, and it became a republic in 1847.

An Incredible Trek

Pan American had their seaplane operation at a place called Fish Lake in Liberia. They had trained local waiters for their eating facility. The waiters were so proud of their jobs and their uniforms with white gloves, that, when land-plane operations were removed to Monrovia, they appeared at the compound many weeks later, ready to work. They had walked a vast distance through unfamiliar territory to regain these prestigious jobs.

Mrs. Johnson, descendent of American slaves, in Liberia, 1946.

A vine bridge in the jungle near Monrovia, Liberia, 1946.

Holy Reserve

Just after we took off from Leopoldville, Belgian Congo one evening, a warning light indicated a malfunctioning engine. Following procedures, we immediately turned around and prepared for an emergency landing. It could well have been a fire, although it turned out to be a minor malfunction.

Our passengers included a number of priests and nuns, and I remember that their skin was extremely yellow from several years of taking atabrine to ward off disease. We were impressed by the way these folks kept their calm and shared their candies and food with everyone during the stressful time. They were French speaking, so they may not have quite understand what was happening. Nevertheless, their calm was impressive. Americans would have gotten hysterical.

A Northern Exposure

At Stephensville, Newfoundland, I remember a young man working in the hotel who had just returned from seeing a movie about New York City. He was very impressed by the large number of people in Manhattan. To put it in Newfoundland perspective, it was "just like Cornerbrook on Saturday afternoon." (Cornerbrook was the largest village he had ever seen.)

Heavy Drinking

U.S. Customs allowed returning service personnel to declare five fifths of liquor duty-free. Flight crew members were given the same courtesy once a month. In Newfoundland and Labrador, almost everyone utilized their quota, purchased from duty-free stores for the ridiculously low price of $1.65 for a fifth of Canadian Club. Unfortunately, this increased the weight of the airplane load by approximately 2,000 pounds—an amount that I doubt was ever reflected on the weight and balance of the aircraft documents. Fortunately, the remaining distance to be flown was not critical, so no problems were ever blamed on the extra liquor loads being carried.

CHAPTER 10

End of the Navigators

> *Pan American begins commercial service to Asia via Hawaii and starts phasing out professional navigators*

In the summer of 1946, Pan American was opening up commercial service to Shanghai, China. I welcomed the chance to get back to San Francisco and signed up.

In Los Angeles, Pan American leased a very small hotel near the Hollywood Bowl for crew layovers. This particular hotel was chosen because of its proximity to restaurants and recreational facilities. The rooms were small, clean, and air conditioned so we could sleep in the daytime.

However, we soon noticed that cab drivers would smile when we gave them the address. And they all knew exactly where the hotel was without directions. It didn't take us long to figure out that this hotel had an entrepreneurial clientele who rented rooms by the hour, not by the day.

Sadie, Sadie, Feisty Lady

One of our 1946 passengers to Honolulu had to wait for passage to Manila. Journalist Sadie Maple was a very small, pixyish lady in her sixties who worked for the National Geographic magazine. She was eager to see how the Philippines had changed since her pre-war visits.

We ran into her later and learned that she got to Manila, but had been refused admittance because of some things she had written were considered unfavorable to the current government. I'm surprised they dared to challenge her. Sadie Maple looked like she could have convinced them to let her in with flourish of her trusty umbrella.

Murder

We had just leveled off at cruising altitude on a 1947 flight to San Francisco from Honolulu. One of the pilots noticed an article in the Honolulu newspaper about two men wanted for questioning in the death of a woman. She had been honeymooning in Honolulu after a mainland wedding. Now her groom was mysteriously missing, and so was his best man who had accompanied them. Murder was suspected.

Our purser glanced over the pilot's shoulder and spotted the story and photos. He told us there were two passengers acting oddly in the cabin, and they fit the descriptions.

We were over the "high seas," and the captain realized there might be legal problems for the airline. He instructed me to keep detailed and chronological entries in the ship's log of whatever happened during the flight. The purser was to report any unusual incidents to the cockpit, and I would record them.

It was not long before the two men began arguing loudly with each other. In the cockpit, we were very concerned. If we tried to restrain them, would they sue the airline? But what if they injured other passengers? Or worse? Before we could reach a decision, several male passengers decided for us. They separated the two men and tied them up. The suspects were stowed in the rest rooms and delivered to the waiting authorities at San Francisco Airport.

No Palms on the Wallpaper

A coin-toss put two middle-aged auto dealers on one of our flights from Los Angeles to Honolulu. They had had a good day at the Santa Anita Racetrack (and in its bar), so they tossed to decide whether to spend their winnings in New York or Hawaii. Hawaii won. They had not told their families or offices about their intentions, and had no hotel reservations.

Their drinking continued on the Clipper. When they couldn't get a hotel room, someone kindly authorized them to use an empty crew room. There they stayed and continued their drinking. They saw only those four walls during their vacation in the islands. When they left, they said they didn't think Hawaii was anything special.

Ethel's Ashes

Navigator David Linhart was asked to take the ashes of a friend and drop them somewhere over the sea on our way to Hawaii. After he gained flight altitude and the crew had settled down for the long night flight, Dave carefully moved the radio operator's feet to one side so that he could open the drift sight hatch in the floor. Underneath the door was a four-inch pipe through which the navigator could eject a small magnesium flare for drift sightings. Unfortunately, the container holding the ashes was too large to fit the ejection chute.

The only thing to do was to open the container, remove the sack containing the ashes, and force it into the chute. But— oops—the sack broke. With predictable results. A cloud of dust swirled through the enclosed flight deck. "What on earth is that?" someone cried. "Oh, it's just Ethel," said Linhart. Not everyone thought it was funny.

Gooney Birds

No account of the Pacific Islands would be complete without mentioning the gooney birds on Midway Island. They are large albatrosses with a wingspread too expansive to permit the standard flapping for takeoff. Instead, they extend their wings and run until they gain enough speed to fly. This isn't instinct. They have to learn to run upwind for takeoff and landing like an airplane does, and to lower their "landing gear" when alighting. Failure to do so results in a comical topsy-turvy crash landing.

These birds perform lengthy ritual dances, and lay their large eggs on unprotected places such as runways and other high traffic areas. They are truly gooney.

Beginning of the End

By now, Pan American was considering the use of more electronic navigation aids. Their plan was to train pilots to take over navigation. Some navigators went back to school to train for new careers in other fields. Others left to start businesses or pursue other activities, some while they were still on the payroll. Even though Pan Am was finding jobs overseas for Americans, it didn't take them too long to realize they could train and use local people for a lot less money. I knew We'd be displaced and decided to be prepared. I started looking around for a business. But doing what?

The Dolls in My Life

Even after the war, many manufacturing materials were still rationed or in short supply. This made it hard on small businesses. However, I discovered that everything needed for making ceramics was readily available. Lots of ceramic companies started at this time.

In 1947, I went to see the head of the Ceramics Department at the College of Arts & Crafts in Oakland and said, "I'd like to hire one of your best graduates. I'll get a building, equipment, chemicals, and pottery wheels. Maybe this woman can come up with some designs, and we can sell them." (Graduates were predominantly women, since most young men had been in the military or defense work.) Eventually I hired two women who turned out handsome ceramic art pieces that sold to top stores like Gumps.

However—and you entrepreneurs will nod understand-
ingly—problems soon developed. The ceramists would make
something truly beautiful, I'd get orders, but then they couldn't
assembly-line it. House and Garden magazine featured one of
the designs, but we couldn't replicate it in quantity. The artists
were college-trained, but not production-oriented. Everything
they made was essentially one of a kind.

After a year, I had three women on salary, but I wasn't
selling anything. They were having a wonderful time being
creative, and I was going broke. I realized this was a no-win
situation. I had to come up with a product that ordinary working
people could manufacturer.

Inspiration came in the form of a china doll's head that
my mother had had since childhood. There must be lots of
grandmothers, I thought, who have one such doll but lots of
granddaughters they like to give it to. Maybe they'd buy
reproductions. I made a mold of the doll's head and got an artist
to design some coordinated arms and legs.

My first small ad ran in a little publication called
Profitable Hobbies. I offered a completely dressed "Grandma's
Dolly" for $5.95, or undressed with dress pattern for $4.30. For
women who could sew, I offered the china head, arms and legs
plus body and dress patterns for $2.70. Postpaid! The ad cost me
$27 and we grossed about $1,000 in a month. Then I found
another publication, The Work Basket, a thin pulp magazine that
went out to one million ladies who liked to tat and quilt, mostly
farm and church women. An ad cost $300 and drew literally
bushel-baskets of responses. Eventually we advertised in Sunset
Magazine, Better Homes & Gardens, and House & Gardens.

In 1948, Pan Am let us go, and I concentrated full time
on the doll business. By October, 1949, we had so many orders
coming in that I had to lease a new building, buy a few more
kilns, and hire more people. Soon customers were asking for
different colors of hair and different sizes of doll, plus salt and
pepper shakers. They offered all kinds of suggestions. Without
really planning to, I had become a doll business, supplying doll

hospitals and producing sixty-four-page catalogs mailed to 100,000 people.

We named the first model "the Jenny-June doll," but the doll-collecting world referred to it as the Mark Farmer doll. In recent years, I've seen some of my dolls in antique shops.

Lots of doll collectors like glass cases, so I soon got into that business. We also carried dozens of doll books and doll stands. I was one of the first to import Hakata dolls from Japan, and I had Alaskan Eskimos sending me fur and ivory dolls. Through a connection with a Chinese importer in San Francisco, we carried a large selection of Chinese dolls. All this time, I was using my flying to make doll connections all over the world. Eventually we had twenty-two employees.

No Navigators Wanted

The Civil Aeronautics Administration (predecessor to the FAA) had decided in 1947 that navigators should be licensed if they were to fly over water. It seemed ridiculous after all those years that we should have to take a several-day exam to prove we were capable of doing our jobs. However, everyone passed with flying colors, and we got the first navigation licenses issued.

Then, in 1948, all Pan American navigators were terminated, and their jobs given to trainee pilots. Thus ended what I thought were my flying days. Because the Navy had never called us to active duty, we had earned no GI Bill benefits and were never acknowledged by the military as having participated.

The seaplanes were gone. Many airlines were flying worldwide. It was no longer anything special. It's the doll business for me," I thought. But then the Korean War started and Pan Am wanted us back.

CHAPTER 11

Rise of the Non-Skeds

After World War II, the public clamors for access to inexpensive air travel, and non-scheduled airlines spring up to meet the demand. They are able to play a key role in the Korean Airlift in the early 1950s.

War brings big changes. Early air travel had always been for the elite and affluent. In the 1930s, a one-way ticket on the China Clipper between San Francisco and Pearl Harbor was $720—the cost of a sporty car complete with spot lights and radio. (Today, $720 would buy a couple of optional auto add-ons.)

Suddenly, it was no longer unpatriotic to travel. The public was fed up with restrictions and positively bursting to see exciting new places. Farm boys had become experienced world travelers while in the military, so foreign excursions now seemed possible.

The "non-sked" operators rushed to respond to this new market. With their minimal overhead and low cost equipment and personnel, they could offer low fares to anywhere. Of course, passengers put up with crowded seating and minimal or no food service, but they got where they wanted to, quickly and cheaply. The non-skeds called this new type of service "Coach Class," a twentieth-century version of steerage on passenger ships. Suddenly, the traditional airlines had to offer similar options so they could compete with upstarts like Seaboard and Western, United States Overseas Airways (U.S.O.A.), Los Angeles Air Service, California Eastern, and Transocean. The old, luxurious service was now called "First Class."

Cargo hauling underwent a similar overhaul. The war had gotten people used to the idea that you could routinely ship goods huge distances in a few hours or days.

By 1946, many ex-military pilots jumped at the chance to acquire surplus airplanes that could be converted to civilian uses. Former military personnel had seen how airplanes could supply and sustain huge movements of cargo in areas where no roads existed.

Flying Tiger pilots from Chennault's group in China were among the first to suggest that an all-cargo airline should be established for the continental United States. The "Tigers" personnel pooled their funds, acquired surplus aircraft and the rest is history. Others started airline service with just one DC-3 or DC-4. Surplus C-54s (DC-4s) and C-47s (DC-3s) were available cheap.

There was a shortage of airplane capacity between San Francisco and New York. Imagine the delight of passengers who had been having a hard time finding a seat on an established airline that would go where they wanted to go. Suddenly, for half the price, a non-sked airline would take them to their destination. Of course, the departure might be delayed until there were enough passengers to make the flight profitable. But, for many, this was outweighed by the economy and convenience.

As Henry Kaiser, the World War II shipbuilding genius, said, "Find a need and fill it." These new young airline operators did not have a staff of experts and advisors on their payroll. Their philosophy was simple: Choose a name that sounds safe and established, paint the name boldly on the fuselage and tail, find a route or segment that could use some competition, and you're ready to go—now!

Ed's Airline

I was having lunch with a Pan American technical engineering executive in the fall of 1946, when the public address system announced the immediate departure of Trans Luxury for New York. Ed Taber, the owner and copilot, spotted me and came over to say hello. The Pan American pilot said, "Mr. Taber, how much does it cost you per mile to fly your aircraft?"

Ed, who was well under thirty, smiled and said, "Piles, Captain, piles." Obviously he didn't know, and couldn't care less. Ed continued, "I'd like to stick around and visit, but I've got to sell a couple more tickets before we can depart."

This was the pattern of many non-skeds. Fill up with fuel, keep your fingers crossed that you wouldn't have any problems, maybe last a few months, and change the name of the airline if a bad accident occurred. (They did!).

When Ed had gone, the Pan American Captain asked me, "Who is that fellow?"

I replied, "Ed Taber. He flew much of the war for you as a Second Radio Operator." (Ed's airline eventually lost a plane, but he wasn't flying it at the time.)

Over the next few years, the "fly by night" operators were eliminated. Those who had a more businesslike approach were able to continue operating.

The Korean Airlift

When the Korean conflict began in 1950, there was once again an immediate need for a Pacific airlift from the west coast. The military gave contracts to virtually all carriers who had equipment, usually DC-4s. This finally gave the so called non-skeds their chance to operate scheduled service overseas with adequate cash flow and a measure of respectability.

Experienced personnel poured out of the woodwork to fill the need for both the scheduled and non-sked airlines. Ex-military pilots, navigators, and radio operators flocked to these start-up operations for another "go."

Shopkeepers, tradesmen, farmers, and salesmen were eager and willing to make a flight a month to Tokyo to augment their income and renew World War II friendships. Whenever an airline needed personnel, an informal informational network got the word out. Crews were recruited by word-of-mouth almost overnight.

Costs of retraining were minimal. There were no job guarantees, no pensions. Crews hoped the job might lead to something better. Where the financing and business methods of the airline were solid, there was success and growth.

The Korean Airlift flew mostly from Travis Air Force Base to Tokyo via Honolulu and Wake Island. Some flights were via the Philippines, Guam, and Taiwan. From those points on, the military did the actual flying to South Korea.

Once again, Pan American offered me employment as a navigator, but I could see no future with them. I declined.

Overseas National Airways

George Tompkins, a former naval officer, had four DC-4s on lease from the government. He re-leased them to Transocean Airlines, formed by Orvis Nelson. Nelson was a former United Airlines pilot who had managed United's World War II Army Transport Command contracts. He was now doing charter and limited scheduled flying transcontinental.

When the Korean Airlift began, Tompkins retrieved his aircraft and started Overseas National Airways (ONA). Its personnel were recruited from former Air Force pilots and those who had served in NATS (Naval Air Transport Service) and ATC in World War II.

I was asked to fly with ONA as navigator, but I hesitated, concerned that it might not be up to the caliber of Pan American's training and maintenance. For several months, I remained on the side lines.

Then, in 1951, I was again asked to join them. If necessary, Tomkins said, I could have time off for my doll business during the busy season. I requested a refresher check ride to Tokyo via Honolulu and Wake Island.

Trips took about ten days with seventy hours of flying time, which meant I would get two weeks off each month. The pay was $450 a month, plus $10 a day for expenses. Compared to the opulence and glamour of Pan American, ONA was a new ball game.

At ONA, as at all the non-skeds, the navigator had to make his own flight plan from information on Air Force weather maps. Altitudes were chosen that offered the most favorable winds—or the least unfavorable winds. Reserve fuel was carried so we could go to an alternate site if weather prevented landing at our destination. However, some atolls like Wake Island were so remote that no alternate airfield was available. (Employees of the airline, oil company, and Federal Aviation Administration on Wake Island had to fly 2,000 miles to Honolulu to go to the dentist.) For such remote spots, we carried several extra hours of fuel to allow for a longer holding time. I suspect there were times when unrecorded "Granny gas" was carried that exceeded legal limits.

To achieve maximum load, seating was crowded. Box lunches were served to military passengers, many of them teenagers. The food was so dry that the crew couldn't eat it, but the kids chowed down and thought it was great. Some of these kids were making their first flights, and many did not realize exactly where they were going. "What ocean is that?" they'd ask, staring out at the Pacific. But the military passengers were always great guys. There were never any disciplinary problems aboard.

ONA had been quickly thrown together with surplus airplanes and no thought of permanence. Operations were from old World War II wooden hangars that had been hastily built a

ONA's DC-6, Baltimore, 1958.

decade before at Oakland Airport Contractors did the maintenance. There were minimum technical and office staffs and crews to do the flying. I considered this work just extra income.

Although ONA was one of the "go! go!" airlines, their safety record was remarkable. They had no operational losses over water. There was one incident when an ONA airplane collided with a Cal-Eastern plane directly over Oakland Airport while both were doing training. The Cal Eastern plane landed in San Francisco with no casualties, but the ONA pilots were killed. I was called in to stay with the deceased Captain's wife until relatives could arrive at their home. This was grim, and the closest I ever came to tragedy in my thirty-eight years in the business.

A National Resource in Emergencies

All airlines, both scheduled and non-sked, are a tremendous backup for quick mobilization in national emergencies. They allow the government to keep some military air power in reserve, never using it at one-hundred percent capacity. By using commercial airlines for temporary assignments, the military keeps its own planes as backup in case of attack or disaster.

Of course, the U.S. military has the authority to confiscate commercial aircraft in emergencies. It's rarely done, but the military has planned for it. Many aircraft are built to be converted to airlift mode in a matter of hours.

The post-war airlines operated at maximum efficiency. Crews could be hired a month at a time with no benefits other than a paycheck (far from punctual in some cases). If one non-sked furloughed its personnel due to a slow month, the crews merely went to work for the rival operator who had caused their unemployment by lowering his fares. In the course of a year, it was not unusual to work for three airlines. We had a saying, "Be kind to the new hire. If he's furloughed before you, he'll be ahead of you at the next airline—maybe as your boss." During the Korean war, I also navigated for California Eastern and The Flying Tigers, but I always went back to ONA where my seniority had accrued.

When the Korean conflict continued far longer than anyone expected, newer equipment was introduced. Monthly bidding for government contracts kept the government costs down. The more marginal carriers were eliminated.

Feast or famine became a way of life. The Korean civilian airlift (which actually went only to Tokyo) was never acknowledged. As far as the public was concerned, the military was doing all the work, and we were nowhere in sight. But we had an amazing safety record to our credit.

When the Desert Storm airlift to Saudi Arabia was on television in 1992, I noted with nostalgia that the markings on the 747 jets were those of civilian airlines.

Wake Island Typhoon

In September of 1952, I traded flights with another navigator so I could see an uncle from Tulsa. The flight I was to be on was at Wake Island on September 16 when typhoon Olive struck with 170 mile per hour winds. The pilots on Wake had to keep the engines running on the ground during this storm so the planes wouldn't be blown off the island. Virtually everything on

Wake was destroyed except some tanks and a concrete food bunker where personnel took refuge.

I arrived two days later. Happily, no one was seriously injured, and no aircraft were damaged. Wake is only a few feet above sea level. The waves crested over the island during the storm. Afterwards, the army flew in a hundred tents so people had a place to stay.

Love in a Box

Part of the cargo of most flights was mail destined for military post offices in the far east. On a trip from Wake to Tokyo, I noticed a fragile-looking box on a crew bunk. It was an ordinary bakery cake box, secured with twine. Postage stamps had been stuck in one corner, and there was an address just below the window. On its side was "Do Not Turn Over."

Inside the box was a birthday cake in perfect condition. A mother in the Midwest was sending it to her son, an Army private in Korea. The cake decorator had carefully inscribed her birthday message in colored frosting. By the time the cake got to our plane, dozens of people had added their signatures. I'd like to have followed that cake. I have no doubt that it reached its destination in perfect condition.

Transporting News Film

In Tokyo, it was not unusual for newsreel cameramen and war correspondents to ask permission to stow exposed battlefield films in the DC-4 cockpit for quicker transport back to the states. There was some foot room below the navigation table, just the size needed. A courier would meet the plane in Honolulu, retrieve the film for Customs inspection, and put it back on the plane for the continuing flight to Travis Air Base in California. It was amusing that I had my feet resting for hours on what would be the "News of the Day" tomorrow, but which was already two days old.

Clark Field

The big American air base in the Philippines was Clark Field. It was about fifty miles north of Manila, and had been the military home of both General MacArthur and a younger Eisenhower before World War II. The base was well established, with old artillery pieces and a parade grounds surrounded with mature trees.

The country is rich in natural resources, and, under the right circumstances, should be prosperous. The military had always maintained a benevolent attitude toward the Philippines. So when a Clark Field fire engine with blaring sirens sped past the security gates one day, nothing was done to stop it. To the best of my knowledge it was never recovered. Some village needed it more than the U.S. military.

Further north and high in the mountains, the military had established Camp John Hay, a beautiful recreation facility in Baguio. Before air conditioning, Baguio offered relief from the tropical heat of Manila and Clark, so it became known as the summer capital of the Philippines.

When we transited Clark Field, we ate at the Officer's Club. The food was excellent and the entertainment was outstanding. A large Filipino band, formally attired, played throughout the dinner hours. They had great talent, knew all of the latest American hits, and were paid wages commensurate with the local economy.

Just to the west of Clark Field and adjacent to a border fence lived a tribe of Negrito Pygmy natives. It was said that they had lived there, on the slopes of Mount Pinatubo, for over 50,000 years. My first sight of them was a woman with very long nipples on her breasts. She was nursing a piglet, something they considered quite natural and practical.

The air field was on land that the Pygmies had formerly held, possibly one of the reasons they were given special treatment. Another was that they had been very effective against the Japanese occupiers during World War II. They rescued downed U.S. airmen and hid them. General Douglas Mac Arthur publicly thanked them for their bravery and ingenuity. After the war, they sold handmade knives and blow-guns as souvenirs of their traditional existence, now giving way to "civilization."

During the Vietnam war, these little people trained Air Force crews in jungle survival skills. They were highly effective as guards and could remain hidden high in the huge trees.

Mount Pinatubo erupted in June of 1991, covering the entire area, including Clark Field and neighboring Angeles City with volcanic ash. The air base was abandoned. However, Filipino friends told me that the Negritos survived and continue to live on the slopes of Pinatubo.

Tokyo Rising

In 1952, Tokyo was still in rough condition. The road into town from Haneda Airport was gutted with pot holes. We stayed in Japanese-style accommodations at the Shiba Park Hotel downtown and later at the Nikatasu, about two blocks from General MacArthur's headquarters in the Dai Ichi Building.

I did a lot of walking around Tokyo. Several times young people, twelve to fourteen years old, would ask if they could walk along to practice their English. They would tell me where they stood in school and list their subjects. Calculus in high school, many sciences, and several languages were the norm. Over the years, I've wondered what happened to these young people. They must have been in the vanguard of the Japanese successes later.

During our layovers in Tokyo, crew members usually ate together in the evenings. Often the local Japanese representative would accompany and guide us to recommended places. Sometimes we would find ourselves late at night in areas off the beaten paths. However, we were always treated with courtesy and honesty. Compared to our American cities, Japanese cities were very safe and law abiding.

Ex-Kamikaze

Flying in and out of Haneda Airport in Tokyo was helped a great deal by having a Japanese representative on the ground who was familiar with details and customs. ONA used a fellow named Keiji who greased our way for years. He was a responsible and well-liked liaison and interpreter. Keiji had been a young Kamikaze "suicide" pilot. Fortunately, the war had ended in time to spare his life. Many years later he was killed in a World Airways accident.

Eye-Aye-Aye

After one particularly grueling flight to Tokyo from Wake Island, our regular hotel rooms were not available. The crew ended up in double rooms in a Japanese hotel underneath a railroad track. We were so exhausted that we never heard the loud overhead rumblings and slept like logs.

My roommate was the copilot, a man who had lost the vision in one eye during the war. I'll never forget waking up the next morning and seeing his glass eye vibrating wildly in a tumbler of water.

The Imamura Connection

Being in the doll manufacturing business and going to the Orient every month, I had had good opportunities to make some importing connections. These monthly flights to the Orient were a source of extra money. I contacted the Bank of America in Tokyo for the name of some reliable Japanese firms who exported dolls, display cases, wigs, and doll stands. They referred me to Mr. and Mrs. Imamura.

The Imamura shop was in a second floor loft in an alley a block off the Ginza, near the Takashimaya Department Store. My translator communicated my desire, and they provided samples. A steady flow of orders ensued. We advertised their products in our sixty-four-page catalog which reached 100,000 doll collectors annually. I was able to see the Imamuras on each monthly trip.

I never experienced any breakage, shortage, or problems with these wonderful people. Their shipments were always correct, and the packing was superb. No torn newspaper or alfalfa. Each article was wrapped and presented as a splendid present.

When I first met the Imamuras, they were probably in their early fifties. They were badly in need of dental care and still struggling to recover from the war. Over the years, I continued to drop in on them. On my last visit, twenty years later during the Vietnam airlift and long after the end of the doll business, I was pleased to note that they now owned the entire building and had beautiful teeth. A shiny black sedan sat in their car port. I had been sure their hard work, conscientiousness, and integrity would pay off, and it did.

The Earthquake

I was in the Imamuras' second-floor office one day in 1956 when an earthquake hit. The jolt was so violent that I had a hard time staying on my feet. It lasted only a few seconds. No one else seemed the least excited, and there was no damage. The next day, newspapers reported that the only person injured was a fry cook burned by sloshing grease.

Much of Tokyo is built on landfill which was once part of Tokyo Bay. Many older buildings were built to float on the landfill like a large boat. The result was that earthquakes rocked the boat without doing much damage. If a structure tipped a bit, hydraulic jacks could straighten it.

Later, a Braniff pilot told me how he was making his approach to the Lima airport at the exact moment a major earthquake hit the area. The rippling of the runway and surrounding land was observed from the air for the first time.

What's Tokyo Doing *There*?

On a flight from Wake to Tokyo one night, my calculations showed that we had over an hour's flying to our destination. Suddenly the lights of a large city lay directly ahead.

Landing your aircraft at its estimated time of arrival (ETA) was an unofficial measure of the navigator's accuracy. Although there were many uncontrollable variables, other crew members would rib you if the arrival time varied too much.

But before I could get too embarrassed, we realized that we were seeing the floodlights of a large Japanese fishing fleet, operating at night.

Conspiracy on Iwo Jima

One evening, on a ferry flight to Tokyo, our aircraft diverted into Iwo Jima. This small island had been the scene of fierce battles near the end of World War II. It is best remembered for the photograph of the Marines raising the U.S. flag on top of Mount Suribachi. Now it was in caretaker status, with little air traffic to break the monotony. Japanese were on the island to perform the proper rituals for their fallen warriors and to salvage scrap metal from the battle fields. Surplus American landing craft were used to transport this scrap metal to Japan.

We landed to refuel. Our captain reported to military operations that we planned to continue immediately as soon as we could get something to eat. All of us, including three attractive hostesses, were driven to the Officer's Club for dinner.

As soon as we reported back to Operations, their meteorologist told us we'd face nearly impossible weather conditions if we took off. Fortunately, he continued, they could provide accommodations for the young ladies, while the male crew members could go to the Bachelor Officers Quarters to await better flying conditions. A most generous offer.

However, the skies were clear and a full moon shown above us. Any delay meant our airplane would miss a scheduled departure out of Tokyo. We realized that if we took off at once, we wouldn't encounter this "impossible" weather immediately. We decided to go for it. We could always turn around if conditions became dangerous.

The plane had perfect weather all the way to Tokyo. We could even see some of the islands below in the moonlight. It was then that we realized that this same moon and perfect weather were also being enjoyed on Iwo Jima. We had spoiled a carefully planned party.

Following the Wagon Ruts

Captain Silas Miller was a Kansas cattleman who hated leaving the responsibilities of his ranch when flight duty called. He often educated us on the hazards of cattle bloating and other bovine afflictions. Tall, athletic, and physically capable, he had been a training pilot for Pan Am, and was finishing his career as a pilot for Overseas National on the Korean Airlift. On long walks during Wake Island layovers, he told me about his years growing up in Dodge City. Si was knowledgeable about Indian battlefields and had the arrowheads to prove it.

From the air, Si could point out the still-visible east-west covered wagon trails etched in the prairie around Dodge City. Invisible from nearby highways, these worn ruts quietly bear witness to an incredible period of our history. No navigation problems for passing aircraft. Just follow the ruts west to Denver.

Following the Contrails

My wife and I were enjoying breakfast one morning in 1995, sitting on our deck in the Napa Valley wine country with our friend, folk singer Glenn Yarbrough. Glenn has twice sailed alone around the world, so he and I share an interest in navigation incidents. I commented that, because of the great amount of air traffic, it would now be possible to navigate the ocean to Hawaii or Europe by following the contrails of airliners.

Yarbrough replied that he had once run into some people who apparently relied on my system. He was at midpoint between San Francisco and Hawaii when he came across another boat that wanted to know their exact location. Glenn was using a satellite navigation system, and couldn't understand their foolhardiness in attempting the trip so unprepared. They knew they were near their intended course due to the contrails overhead, but had no idea what their progress was!

Bonsai!

Small potted Japanese Bonsai trees have fascinated me for years. I bought a particularly beautiful one at the Takashimaya department store on the Ginza in Tokyo. Knowing that U.S. Department of Agricultural inspection at Honolulu would not permit entry of Japanese soil, and that the tree would have to be fumigated in Honolulu, I instructed the store to remove the soil from the tree, then pack the container in a separate box and wrap the tree in moist tissue for the trip.

All went well with the agricultural inspection. But when I got to my hotel in Honolulu, I found that the soil had indeed been removed from the tree's container—but included in another carefully wrapped package.

Time with the Tiger

In the fall of 1956, Overseas National "furloughed" a lot of employees. I was one of them. I knew I'd be recalled soon, but I applied for work at the Flying Tigers Air Line in Burbank. They hired me at once, and I spent the next day updating my passport visas for Japan and the Philippines.

To my astonishment, I found that the Tigers were being sent to Japan to live. From there they would make the Tokyo-Okinawa-Taipei-Clark Field run that I had done for ONA. Some of the owner-pilots were from General Chennault's original American Volunteer Group (AVG) in China. I learned that many had investments in Japan and were doing quite well.

Flying Tigers' paper work was very informal, probably to be expected from a soldier-of-fortune group who had seen real combat under primitive conditions in China. About six weeks later, ONA did recall me. I didn't even get my uniform until after I had left.

Rather than returning to San Francisco, I offered to go to Wake Island and stand by and take over on ONA's first aircraft coming out. This saved their having to deadhead a navigator to Wake.

Hostess with the Mostest

Throughout the Korean war and even into the Vietnam conflict, military airlift was needed between Clark Field in Manila and Taipei, Okinawa, and Tokyo. World Airways was awarded one of these contracts for their very low bid.

World used Japanese women as cabin attendants, a considerable savings because they could be based in Tokyo, saving the usual per-diem costs for Americans in Japan. These hostesses were eager to please and had the additional advantage of not needing to be rotated home to California. Their training emphasized taking good care of the passengers.

One young hostess did not report for her second flight out of Tokyo. It was later discovered that she did not realize that her obligation to her passengers ended when the flight had reached its destination.

The Tidal Wave

A tsunami warning had been issued in Honolulu. A tidal wave was expected. Our crew, resting on Waikiki beach, had no idea just what to expect. Would a violent surge of water inundate the city? It might be hours before the phenomenon arrived—if it did.

While we waited, I took a walk toward Diamond Head. Abeam of Kuhio Beach, I noticed a slow, gentle receding of the water, as though someone had removed a drain plug. Within a half hour, the ocean floor was exposed for several hundred yards out to sea. Marine life was left floundering. There was nothing violent, just a slow withdrawal of the water, followed by an equally leisurely return.

I learned later that the wide harbor had lessened the effect of the tsunami. Areas with a narrower, v-shaped harbor are most likely to suffer because the wave builds speed and volume as it rushes through the bottleneck of the inlet.

Un-Christmasy

In the mid 1950s, ONA flew several DC-4 charters from Honolulu to Christmas Island, 1400 miles due south of Hawaii. Our planes were delivering personnel and material for the British atomic bomb tests. There were no facilities there, so we refueled and departed immediately. This whole operation lacked the personal satisfaction of an evacuation or mercy airlift. Not particularly joyous.

Tanned All Over

On Lewers Street, near Waikiki, a number of budget-type hotels had been built soon after World War II. They were popular with Australian vacationers and were often used by airlines for crew rests. On one occasion, word got around that if you watched a certain hotel window at four o'clock every day, an Australian school teacher would appear on the balcony of the hotel across the street, wearing considerably less than a bathing suit. By counting windows and knowing the floor level, her room number could be computed. Soon her phone would ring with invitations for dinner.

This lasted for several weeks, presumably until her vacation was over, and she returned home.

A Higher Authority

We were inbound to the west coast on a cargo-only plane. The weather had been difficult, the winds adverse, and if the trend continued, there could be a fuel problem. Earl, the copilot, was becoming increasingly nervous. He was a big man in his sixties, a former Civil Aeronautics Authority (CAA) inspector, balding and usually jolly. Although copilots shouldn't leave the cockpit during such conditions, Earl suddenly got up and strode aft.

He was gone about ten minutes. When he returned, he seemed greatly relieved and reassured. He told us he'd had a talk with the "man upstairs" about the safety of the flight and everything would go well. It did.

Many Hats, Just One Head

On one flight over the north Pacific, we stopped at Cold Bay, Alaska to refuel. Cold Bay is in the Aleutian chain of islands. It was obvious there was little traffic through this area. This was our point of entry into the United States, so we had to clear U.S. Immigration and Customs.

A heavy-set woman in her forties acted as the local representative of both bureaus. After she cleared us, we noticed a Reeves Aleutian Airline aircraft landing. The woman informed us that "they must be bringing the prisoner in. I'm also the judge here." No doubt she was also Postmaster and the whole government wrapped up in one.

Playing the Charter Game

As the Korean War wound down, non-sked operators began looking for ways to stay in the game. Charters, both military and civilian, seemed to be the answer.

However, the major airlines were complaining so loudly about lost business that new regulations were passed. The government severely limited the number of flights a non-sked was allowed to make to Europe each month, making profit nearly impossible. However, the government agreed to grant one-time permissions for individual flights chartered by organizations. If you could fill a plane with affiliated people, you could fly anywhere in the world.

A loophole! Travel agents began offering "club membership" to anyone and everyone flying on a particular plane.

European charters were very popular during the summer, but to organize them and obtain government approval still required months of painstaking planning. It took creativity to deliver a load of passengers to somewhere in Europe and then have a plane available to bring them home several weeks later without flying any empty planes. Charter airlines had to be sure that fuel and ground handling facilities were available at places they'd never been. Some stops even required cash payment. There was one incident where the pilot had to borrow a passenger's gas credit card to refuel the airplane! Catering to passenger requirements also required some ingenuity. Sometimes Kosher food had to be procured in places that had never heard of it.

Still, charters were the way to go, and each carrier fought for its share. Sometimes, by comparing notes, we would learn that a job we were sure was ours had been double-booked with another non-sked. It was a highly competitive world.

Eased into Management

In 1957, the city of Baltimore had just completed Friendship Airport north of Washington, D.C. It was designed to be the pride of Baltimore and to take some of the pressure off Washington National Airport. There were handsome new offices and maintenance facilities, not yet utilized to any extent.

George Tompkins, CEO of Overseas National, was invited to move his headquarters from Oakland, California to Friendship and become Baltimore's own airline. We had just taken delivery of the last four DC-6s from Douglas Aircraft and now had brand new equipment. Sales offices were maintained in Washington and New York, while all of flight operations moved to Idlewild Airport in New York City.

At this point, I was being eased into management in case any of the numerous projects jelled. We had just finished operating what we called the Missile Lift Contract, providing daily flights from factories manufacturing components for the Cape Canaveral rocket launching project. A DC-4 and DC-6 serviced San Diego, Los Angeles, and Denver, with daily pickups of engineers and equipment for the tests. By having a more east coast presence in Washington and Baltimore, ONA hoped for more government business and European charters.

Finding Help in Hard Places

The Overseas National Pacific-Korean airlift had been a military operation, using military airfields and support facilities. When ONA and other small airlines converted to world-wide civilian charter operations, they were strictly on their own. This meant we needed to find airports that would accept us and arrange for mechanical services, refueling, catering, and passenger handling. All this had to be accomplished without having any company personnel stationed abroad.

Every departing pilot had to be given a detailed itinerary of where to go and whom to call if they needed assistance. Larger airlines with fixed routes and schedules already had personnel and communications at every airport.

Herb Fixler

Herb Fixler became "Operations Manager" at Idlewild Airport in New York when the company moved there from Oakland. He was a small, bulldog type of New Yorker who had been a reporter on a Long Island newspaper. His navigation experience was with the Army Air Corps during World War II. Herb was very intelligent and able to keep details in his head. Without any staff, he was able to get the 1957 charter season started. It was a tremendous, fast-moving effort, done by one man with teletype and telephone. Herb was on the job seven days a week. When the airport was snowbound, he didn't even try to go home. His frequent headaches were understandable as he kept all aircraft and crew movements in his head because there was never time to capture the ever-changing picture on paper.

I was asked to come to New York to assist Herb and give him some prospect of relief. He managed to survive until the season slowed and he could take some time off.

When ONA was awarded a two-ocean contract in 1959, Herb joined me in California to do our crew scheduling with far less pressing responsibilities. He was obviously tapering off, but continued to have those debilitating headaches. Sadly Herb died several years later from a massive brain tumor. His dedication and genius went unrewarded. I learned some valuable lessons from him and felt honored that he thought I could help.

Shooting Down the Fly-by-Nights

During this period, a promoter came into the Washington sales office. He needed an airplane to move a number of loads between Hawaii and the mainland in a hurry. It was "check in advance," and the plane flew vacationers to Honolulu immediately. The check was deposited in a D.C. bank on Friday and bounced on Tuesday after we had flown the first round trip.

This fellow had been offering ridiculously low fares to Hawaii to purchasers of Israeli Bonds. He was obviously losing money, so he decided to let ONA share his expenses. When ONA cut him off, there were still passengers stranded, and the press was quick to criticize the airline. It makes one speculate about the final disposition of the bond money.

ONA Unionizes

All bids for charters were competitive, with the lowest bidder getting the business. The losers furloughed their personnel until they could sharpen their pencils and drop their prices, or until demand took an upswing.

Most supplemental airlines were undercapitalized, and could count only on summer vacation travel or military contracts. It was difficult for crews to make any sort of family plans whatsoever. There was no job security, even for the most senior people, and frequent relocating meant moving children from school to school. The outlook for a normal family life was bleak.

The pilots felt that a possible solution would be to join the Airline Pilots Association (ALPA). But from a management standpoint, the union restrictions and higher pay scale would make ONA operations impossible. I recall the ALPA negotiator saying, "If you cannot afford the pay scale and union working conditions, you should not be in business." He said this in front of pilots who couldn't get employment with more stable companies and who desperately needed their jobs.

For all practical purposes, a young pilot "marries" his company for the duration of his career. Pilot positions are earned by seniority within the airline. Experience with another airline doesn't count. It was unlikely that an experienced and thoroughly qualified captain would happily start over at the bottom of someone's seniority list, even if a job were offered.

ONA had no alternative. It signed with the union, and, ultimately, the pilots lost their jobs.

More Money or Else

As equipment got larger and faster, planes became more productive, so there was a tendency for pilots to ask for more and more money. Actually it was the equipment that was more productive, not the pilots. The genius of Boeing and Douglas engineers enabled airlines to expand and pilots to upgrade to larger aircraft, justified or not. Promotions and pay raises were based on seniority, plus the fact that a work stoppage would shut down a cash-deprived company.

The airline executives negotiating wages were often former pilots who had been active in the Air Line Pilots Association. They related more to their own interests than to those of the company: "If you can't afford to pay what is asked, then you shouldn't be in business." Witness the situations at Pan American, Overseas National, and Braniff.

An expensive game of musical chairs developed as airlines expanded from domestic to international operations and began to operate with a wider variety of equipment, from smaller planes to huge, long-range planes. The larger equipment paid more, and enabled personnel to perform their monthly flying time in fewer days, which made those slots most desirable. Every time an opening occurred on the more desirable runs, the replacement candidate had to be trained in ground school, on flight simulators, and with actual flying. Then there needed to be time for moving the family. Simultaneously, the replacement for the replacement had to go through the same process. And his replacement. This kept things lively. The system was fair to the personnel involved, but was devastatingly expensive for the airlines which were trying to buy new planes for opening new routes into areas that might be questionable economically.

Contrast this situation with a brand new airline operating with only one type of equipment out of a hub that permits more permanent, stable domiciling. Pilots of a failed airline would go to this start-up company at lower wages and were willing to accept fewer airline-restrictive work rules. The new, smaller, more flexible organization with good management is likely to be more profitable.

Pay scales are generally related to an expected standard of living, and expectation is much higher in the U.S. than in most other countries. Pilots of foreign airlines are probably as skilled as those in the U.S., and they have job security because their governments usually see that the airline remains in business, if for no other reason than national prestige.

To be competitive with foreign airlines, an American company operates with a smaller margin of profit, no government guarantees, and more airline-restrictive work rules. The same conditions are true in other fields of production throughout the world. To maintain a level playing field, wages and working conditions ultimately need to be similar everywhere. This is unlikely to happen soon, and is threatening the standards we are used to. Does "level playing field" mean we must all have the same quality of living? If so, we may be in for some disappointment.

The Social Impact of Airline Economics

During the Hungarian revolt of 1956-57, the non-skeds, now called "supplemental carriers," were given charter contracts by the United Nations to fly Hungarian refugees from collection points around Hungary to western Canada and the United States. Pickups were at places like Belgrade, Frankfurt, and Vienna. The problem was that the planes had to ferry (fly empty) to Europe. It was imperative to find a way to generate revenue eastbound.

Well, Jamaica in the tropical Caribbean was a part of the British Commonwealth, so Jamaicans could live and work in Great Britain at any time. The supplemental carriers that were flying empty eastbound decided to offer their seats to impoverished Jamaicans at ridiculously low fares. Travel agents in Kingston generated load after load of ambitious, hardworking people wanting to get to London where they hoped their lives would change for the better.

These flights would stop at Gander, Newfoundland to refuel, change crews, and allow the passengers to get something to eat in the terminal. It was heart wrenching to see young women file off the airplane in the dead of winter, wearing flimsy dresses. They had never seen snow or experienced cold weather, so they had no coats or sweaters.

Arrival in London must have been a real letdown for many of them. For those without nurses' training or teaching certificates, life could be cruel. There were actually pimps who tried to recruit young women in the London terminals.

This mass immigration caused Britain some real social problems—and all because we were trying to help Hungarians and had empty airplanes flying eastbound.

In America's colonial times, sailing ships took raw materials from the colonies to Europe, picked up manufactured goods and sailed to Dakar in Africa where the goods were exchanged for slaves, then sailed to the Caribbean where the slaves were traded for rum. The rum was loaded and the ships sailed back to the colonies to start the cycle all over again. The social impact of cargo economics. Sound familiar? Whenever a large group of people are transferred in one direction, there is an immediate economic need to find some way to pay for the return trip, whether by sailing ship or airplane. But often one problem seems to create another.

When the non-skeds booked charters to the Caribbean, it was monetarily imperative not to fly back to New York with an empty aircraft. Puerto Ricans are American citizens and free to travel to the United States. So, in the 1950s, several hundred thousand Puerto Ricans were able to take advantage of cheap non-sked rates to New York.

At Idlewild (later renamed John F. Kennedy Airport), travel agents would routinely oversell charter flights to San Juan, Puerto Rico. Riots at the check-in counter were predictable as Puerto Ricans with hard-earned tickets for a precious visit home were denied boarding. It was a dangerous situation, and crews always gave the ticket counters a wide berth when going aboard.

A French Stopover

We had delivered a plane to London and were scheduled to layover there, take another airline to Zurich, layover again, then fly to Belgrade, Yugoslavia to pick up our refugees. We always enjoyed London where we usually ate in the small dining rooms of old pubs. But this time we decided to fly immediately to Paris as passengers, spend an evening and day updating memories of previous visits, and then take a slow train to Zurich. This would take us through the French countryside and on into the mountains of Switzerland. It wouldn't cost the airline any more for us to do it this way, so off we went.

With no hotel reservations in Paris, we ended up in an ancient walk-up hostel with primeval bathroom facilities at the end of a hall. That evening, we walked through Pigalle and on down to Place de la Concorde in the few hours available. I thought of my Uncle Clifford who had learned to drive my father's automobile in 1915. While still a teenager, he had volunteered to join the French army as an ambulance driver a year before the Americans entered World War I. Now I was retracing his steps over forty years later. The next day, the slow train gave us an intimate feeling of rural France at its frequent stops.

We reached Zurich in the evening, ate in a small hotel restaurant, then went upstairs to get some sleep before the next day's flight to Belgrade. The beds were old-fashioned feather beds with invitingly thick comforters. I could envision being awakened by a cuckoo clock.

Almost as soon as we got comfortable, the telephone rang. Our plane was early, and we were to report immediately. We flew empty to Belgrade, Yugoslavia, landing under snowy conditions in the Communist-controlled country.

Their New Life

President Tito had been admitting Hungarian refugees who were fleeing their Communist homeland. The United Nations housed them in camps awaiting our arrival to take them to North America. After the plane was loaded and the boarding ramp was pulled away from the aircraft, we noticed one workman still on the airplane. We called for the steps to be put back into position so he could get off. He waved at us to indicate "No, don't bother," and jumped from the plane to the concrete with no apparent ill effects. Yugoslavs were very robust people.

To us, this flight was just another day's work, but to those in the back of the plane a new chapter in their lives was unfolding. They had escaped from Hungary, spent months in refugee camps, and now were crossing the north Atlantic to western Canada. For some it was their first airplane ride. Yet they showed no emotion, probably dulled by all that had happened to them so quickly.

The Soaring Gourmet

We had brought a planeload of Hungarians into Vancouver, British Columbia and were flying empty back to Europe. The head chef of the hotel where we stayed had been trying to book passage to our destination and suggested that, if he could ride with us, he would personally cater the crew with food we would never forget. The normal fare on these Hungarian airlift trips was dry box lunches, consumed only if hunger took over. So, picture us in an empty airplane, being pampered by a white-clad chef with shrimp salad, roast beef, fancy pastries, and fresh-ground coffee. Certainly a win-win situation.

Freezing Kings

During a several-day layover at Glasgow, Scotland, we took a train to Edinburgh to visit the castle. It is high on a good-sized hill overlooking the city below. The sheer immensity of stonework and security planning is still impressive. The huge formal rooms intended for Royal splendor were stonily cold. The fireplaces could not possibly have heated them adequately. As I looked down on the surrounding countryside, I could imagine far more comfortable conditions inside the simple cottages of the common folk.

A Spinning Compass Needle

I got to use celestial observation once again when we were flying a group of refugees from Goose Bay, Labrador to Vancouver, British Columbia in a DC-4. We were so close to the magnetic north pole that the compass became useless for several hours. I had to use the sun compass which meant constant calculation of the sun's true direction bearing from which the true heading of the plane could be set into the gyro compasses. Because gyro compasses wandered, the sun calculations had to be repeated constantly.

Monkey Business

In 1958, monkeys were in big demand for polio research. For many months, there were charter flights bringing monkeys to the U.S. from India and the Philippines. The DC-4s were stripped to cargo configuration to accommodate the cages.

Animal attendants flew in the cabin to tend to the monkeys' needs.

By this time, I was in management and not about to assign myself on one of these very smelly trips. Crews never went aft when there was live cargo, but still their wives grumbled about the odor that clung for days after their husbands returned. Sadly, there were always some monkey deaths en route, but also some births. No one actually knew what the net "passenger" count was. At times, the monkeys would escape their cages and run loose in the hangars, both in Oakland and New York. It was a lot of work catching them, but the guys thought this was funny.

Gun-running and Revolutions

There were several charters from Italy to Cuba during the Cuban revolution. The pilots operating the flights discovered that the cargo was guns for the Batista regime in Cuba. Armed guards accompanied the flights to ensure there was no deviation from plans.

After Castro took over, one of our planes going from New York to Kingston, Jamaica couldn't land because of bad weather. The pilot proceeded to Camaguey, Cuba where he landed safely late in the evening. He telephoned us in New York that the Cubans were suggesting "hoteling" the passengers and wanted cash for fuel and their services.

We didn't want the passengers to spend the night in Cuba. If the government realized who we were, Castro might impound our airplane because the airline had previously helped Batista. There could be some real problems. So it was arranged that Pan American would let us charge fuel on their account. Fortunately, the weather cleared in Kingston, and the flight took off again.

There were also times when small African countries would be at war with their neighbors and, unknowingly, flights were made through these zones by people like us. We might have been shot down.

Remote as the Moon

The Saudi-Arabians purchased a helicopter which ONA agreed to deliver to Jidda. A double crew was utilized so longer segments could be flown without rest stops. (On some long flights it was not practical to have relief crews waiting at the stopovers, or to have passengers layover while the crew rested. In these cases, double or multiple crews were used. One crew would fly while the other crew slept in bunks or sleeping bags. It was grueling, but it got crews home much sooner.)

From a base in Tripoli, we flew through Libya, around the southern part of Egypt, across Sudan, and over the Red Sea to Jidda. We probably did not have permission to overfly Egypt, and maybe did not have clearance from Libya or Sudan. Midway in the trip, we saw a small narrow-gauge train slowly making its way to some wadi that depended on it for contact with the outside world. The desert areas of these countries seemed as isolated and remote as the surface of the moon, and just as strangely beautiful.

Pilgrims

Many small supplemental airline operators went to the Middle East to transport pilgrims to Jidda in Saudi Arabia for the Hajj holy season in Mecca. All Muslims try to make this journey at least once during their lifetime. Often life savings were spent on this journey.

My principal memory of Jidda is that the early morning temperature was 120 degrees. However, temperatures at 3,000 feet could be quite chilly. One pilot reported that passengers had started bonfires in the aisles of the aircraft because they didn't know how to turn the heat on or even that they could request it.

I never made one of these Hajj trips, but those who did described packed misery and exploitation of the pilgrims once they were on the ground. No accommodations awaited the faithful, and they were charged high prices for drinking water.

I Take Charge

In 1958, military charters still operated in the Atlantic and Pacific. American Airlines had purchased a fleet of jets from General Dynamics who had acquired a small fleet of DC-7s as trade ins. General Dynamics set up a new division called General Aircraft Leasing Corp. to find some way to utilize these "obsolete" airplanes which were really not very old. They approached Overseas National, now at Friendship Airport in Baltimore, to see if ONA could use the DC-7s for the fiscal-year military airlift that would begin in October of 1959.

This was to be a one-year contract, going to the lowest capable bidder. Every airline interested in bidding knew exactly how much equipment every other airline had. Most hoped for a piece of the business, which would help sustain them through the slow winter months when tourist charters were off-season.

The rest of the industry thought that Overseas National had only four DC-6s. They didn't know that about twenty DC-7s would be available for military contracts if needed.

In July of 1959, Overseas National was awarded approximately sixty-five percent of the Fiscal-1960 Congressional appropriation for military airlift. There was an immediate industry-wide outcry. Other airlines complained that military contracts were supposed to benefit all airlines and help protect them against unforeseen emergencies. However, the appropriation guidelines didn't say that.

Overseas National was awarded contracts in both Atlantic and Pacific, with service to start October 1, 1959. The Atlantic portion was to be operated with the DC-6s and the Pacific with the newly-acquired DC-7s.

I was given responsibility for the Pacific operation and immediately went to San Francisco to set up operations and find a maintenance facility. Pan American offered us office space and agreed verbally to perform maintenance on the 12 DC-7s. Spare parts were to be cannibalized from extra DC-7s.

Captain Kenneth Healy came along as ONA's Chief Pilot. He had the enormous task of recruiting 300 crew members

Volume 1   March 1960   Number 1

FROM THE EDITOR

As a general rule, editors, like children, should be seen and not heard. Inasmuch, however, as this is the first issue of ONA NEWS we thought we would take this opportunity to welcome our many readers, wherever they are, and whomever they might be. A special word of greeting and a word of thanks to the editorial staff whose efforts, thus far, have assisted in making ONA NEWS a success.

There was a time, and not so long ago, when one employee could, and very often did, know every other employee within the company, on a personal basis. Since our expansion last October, this situation, except in a few cases, has largely disappeared.

The primary purpose behind the inception of ONA NEWS is to alleviate this condition. Through the medium of a company newspaper we hope to bring all the employees of ONA together in a sense, if not all together geographically.

In subsequent editions we will
(see pg. 4, col. 1)

OPERATION PACIFIC

It is not intended that ONA NEWS be a vehicle for presentation of the Company viewpoint. However, it does seem fitting now to comment on an intangible which is so very important to the success of any operation. Time after time the various department heads have observed the speed with which the "new hires" seemed to integrate into the group. Subsequent to any expansion such as we had in the Pacific area, it might well be expected that a dilution of the excellent company/employee relationship might be experienced. Exactly the converse has been true. Virtually everyone seemed to grasp the spirit of the project. The new equipment in this division put "old timers" and "new hires" in ground school together. It was a fresh start for everyone; we were all challenged by the job to be done, and once this spirit became apparent, it came as no surprise to us that the show would go on as scheduled.

MARK FARMER

in less than ninety days. We hired experienced hostesses and trained dozens who had never flown. Pilots, navigators, and flight engineers who had been facing the loss of their jobs jumped at the chance for a year's work. When Transocean Airlines ceased operations, their personnel came aboard.

Then an unexpected setback. Pan American, still chafing over the David-and-Goliath developments, decided to retract their offer of office space and maintenance facilities.

It took me only a day to find new space at Oakland Airport where maintenance would be available, probably due to the coming demise of Transocean. I flew to Tokyo to meet Colonel Herman Rumsey, who would be the Commanding Officer in charge. We obtained maintenance and ground representation for Tachikawa Air Force Base on a contract basis from people we knew were reliable.

At Idlewild Airport in New York, ONA was in need of a Chief Hostess. We put out the word that we were hiring and hoping to bring someone fresh into the operation. As soon as Ellen Clay applied for the job, no one else was even considered. She went to work training hostesses, procuring new uniforms, and prepared for the 1959 summer charter season. My life was about to change.

A Spanish Nightclub in Greenwich Village

When ONA got the Pacific contract and I was to be sent to California, Ellen Clay came into my office to say good-bye and to ask if I'd be able to have a going-away dinner at her father's Greenwich Village nightclub.

Ben Collada's club, El Chico, had been operating for over thirty years, offering Spanish cuisine and entertainment. "El Chico, as Spanish as Spain" had been the site of regular NBC radio broadcasts in the 1940s. Many Hollywood and Broadway producers frequented El Chico and became close friends of Ben Collada. Often, they'd offer him a chance to invest in their shows. He turned down one producer because the proposed show was about "some guy playing a violin on a roof."

Ben's mother lived in the Village. She'd never learned English, and cut quite a swath among her Spanish-speaking neighbors. On one occasion, she was complaining about the high cost of meat and fish. Of course, the restaurant used only the finest cuts and enjoyed prices less than retail. So Ben told his mother to order what she wanted from his sources, and he would see that she got a good deal. When the next bills arrived, there were enormous charges for his mother's selections. She had told all her friends that they could add their orders to hers and enjoy both quality and big savings. Ben discovered he was feeding the entire neighborhood. When he mentioned the huge bill, his mother replied, "Don't meddle in my business."

Ben Collada had a very old parrot, Señor Carr, who spoke excitable and often vulgar Spanish. Public relations people maintained that Señor Carr belonged at one time to Pancho Villa, the Mexican rebel leader. The bird often spoke loudly during the entertainment which could be embarrassing, but it seemed to add to the show. Around four o'clock one morning, several thieves broke into El Chico. They found only a small amount of cash. Suddenly Señor Carr began screaming obscenities and raising a big fuss. This cut short the burglary as the men fled to save themselves from the excitable Spanish "watchman."

Collada made frequent trips to northern Spain and South America to audition and recruit promising entertainers and musicians for his New York club. He was responsible for importing Señor Wences, the talented ventriloquist who produced a series of characters from a small box. Wences had been Collada's boyhood neighbor in Aviles, Spain. For years, when Wences entertained at Ellen's childhood birthday parties, she thought there was another child in Señor Wences' box.

Carmen Miranda, Xavier Cugat, and many well known Flamenco dancers first starred at El Chico. On the evening of my first date with Ellen, Ben told us that Marco, one of his bus boys, was retiring that evening, and there was to be a big party for him.   Marco had spent thirty years at El Chico. He had never learned enough English to become a waiter, but seemed happy to be in the background. He lived frugally in the Spanish section of the Village, saving his money so he could go back to Spain as a "rich man." After the orchestra played a fanfare to bring Marco onto the floor, Ben Collada presented him with a watch, thanked him in Spanish, and wished him a happy retirement. For one brief moment, after thirty years, Marco was in the spotlight.

Several months later, we learned that Marco, who had never learned to drive an automobile in New York, had been killed in Spain while driving his new car back to his village.

Ellen came out to Oakland in October of 1959 to "review the troops" for a week, then returned to New York. We were married in February, 1960.

ONA's Low Bid

The one-year Military Air Transport contract began October 1, 1959. ONA's bid was low because the DC-7s were not going to be expensive, repair parts were available from other parked aircraft, and new hires were less costly. ONA's theory was that our competition would starve while Overseas National had a year of guaranteed business. If we could get a similar contract the following year, there would be no start up or training costs, and perhaps some competitors would have been eliminated.

What actually happened was the government offered additional airlift contracts which were bid at considerably higher rates. Our competition prospered with this additional "Call Business."

Who's Interviewing Whom?

Several weeks after operations began, Executive Vice President Jim Forrest and I were asked to go to Scott Field in Belleville, Illinois to see a General who was monitoring our contract. He received us cordially and afforded us the courtesies given someone with the rank of Colonel. During our tour of the facility, we were told that the General expected us to have dinner with him and his wife at his quarters on the base. After dinner and cordials, the General dismissed the half dozen members of his staff, but indicated that we were to remain. He wanted to compliment us on the operation so far, and to let us know he was "hanging up the blue uniform soon" which made him available for employment. All this lavish treatment because he was angling for a job interview.

High-Stakes "Poker"

Carriers were paid for the number of passengers or amount of cargo they offered to airlift to destination on each flight. If the bid was accepted, they were paid for the agreed total. It didn't matter whether they actually carried less. It was rather like a poker game, figuring the most you could offer to carry without getting caught short.

Because California-to-Honolulu was the longest and most critical leg, we knew that, if we could get to Honolulu, the next shorter distances could be operated without fuel problems. So the navigator would study the weather maps, make out a flight plan calculating the amount of fuel needed, and make an offer of a load to be airlifted. The airline would be paid its offer.

It helped to know the cargo people who were to load your aircraft. If we learned that a load would be 3,000 pounds and our planes could carry 3,000 pounds, we could still submit a price for 10,000 pounds. We could offer the maximum amount and get paid for it, although we couldn't have lifted it anyway.

Les Forden

Every office had to have an administrative assistant secretary who could handle the teletypes and do everything to make the rest of us look good. Jean Paget had been with us previously in Oakland, New York, and Baltimore, and now, fortunately, she was back in Oakland to keep my office in order. In January of 1960, I confided to her that Ellen and I were going to get married in February. She replied that she and Les Forden were also getting married.

Both Les and Jean were in their late forties. Les Forden had been hired to keep office and pilot manuals current for our crew reading-files. He had been around the airport for years doing similar office work. His enthusiasm for aviation was tremendous, but he couldn't become a commercial pilot because he had lost an eye which interfered with his depth-perception. (Yes, one-eyed pilots had been around since Wylie Post, but commercial licenses were more restrictive.)

Shortly after their marriage, Jean told me that Les had cancer of the throat and would not be able to say anything for a year. Would this affect his job? Of course not. For the duration of the contract, Les carried a small pad on which he wrote his messages to all of us. During that year of silence, he researched the history of Oakland Airport and the numerous firsts in aviation that had taken place there.

Les Forden's book, Glory Gamblers, detailed the "Dole Race" from Oakland to Honolulu. Even though Maitland and Hegenberger had flown nonstop between Hawaii and the mainland just a month after Lindbergh's 1927 flight, the Dole pineapple interests wanted to keep people focused on the possibilities of an ongoing air connection. They offered a $25,000 prize for the first plane to make the trip within one year, starting in August of 1927. Art Goebel, the same man who flew my grandmother over Los Angeles, won the money and gained a place in aviation history.

In 1973, Forden published another book, The Ford Air Tours - 1925-1932. These tours proved that planes could be scheduled for cross-country flights, and that passenger-carrying airlines were indeed possible. Forden's book brought back powerful childhood memories. On July 9, 1927, I had boarded a huge (to me) Ford Trimotor as it transited the McIntyre Airport in Tulsa on the Ford tour. The plane never left the ground, but my eight-year-old imagination had soared.

Les Forden turned what might have been a catastrophic year of silence into a career as an historian, documenting events that might otherwise have gone unrecorded.

Green Eggs, With and Without Ham

All during the Korean Airlift, cold box lunches had been served to the military passengers, but our new contract specifications called for hot in-flight meals. This presented a real challenge at Wake Island. A caterer in Los Angeles came up with a novel idea. Why not serve frozen TV-type meals that could be heated in the aircraft galleys? It worked.

Frozen meals were shipped from Los Angeles and stored in large oceanic transport containers with refrigeration units, ready to be loaded onto the transiting aircraft.

On the first trial of frozen-egg breakfasts, the eggs turned green. Corrections were quickly made, and the rest is history. For their day, such meals were quite an improvement, although airline food became the subject of a lot of jokes by a later generation of comedians.

Congress Changes the Rules

Morale was great, the crews were magnificent, and there were no serious incidents in flight. We hoped the contract would be renewed for another year if everyone did a good job for the Air Force.

However, politics began to intervene. Congress had not intended for one carrier to operate that much of the appropriation for airlift. They wanted more airlines to participate, each on a smaller scale, but with the capacity to expand in a future emergency. ONA might be the most efficient carrier with the lowest costs for the taxpayer, but the military really wanted a quicker response capability from many airlines.

The ground rules were changed the next year. There would be no longer be any big contracts. Rather, everyone would participate via smaller fixed contracts.

The Civilian Reserve Air Fleet

The small air carriers were constantly trying to reconfigure their planes so they could offer more seats and lower per-passenger costs on military charters, but not at the expense of compromising comfort or safety. The airlines recognized that the pool of engineering talent that could do this was worth tapping.

The Civilian Reserve Air Fleet (CRAF) was created to give priority to carriers which had cargo doors and could be converted from passenger configuration to cargo within a few hours. Planes were rushed off the assembly line to meet CRAF specifications. It was not unusual to fly a load of cargo to Tokyo with the seats and galleys stowed in the baggage area below, then reinstall the seats within eight hours so we could return as a passenger flight.

Several of the supplemental carrier operators like Kirk Kerkorian and Edward Daly were in better financial shape to survive than other operators, and they did. As newer airplanes became available, it was no longer experience and skill that mattered most. All operators could provide that. Rather, the ability to finance new equipment became all-important. For a time, survivors were those who had good money sources.

The Biggest Load Ever

In 1960, one of our copilots, a young man named Conroy, learned that the government had a problem. Its huge rocket sections were too big to airlift to launch sites. Ground transportation was equally impossible.

Conroy figured out that a Boeing Stratocruiser could be fitted with a huge if ungainly superstructure that could accommodate the rocket sections.

This aircraft was procured. With a minimum of engineering and some very clever mechanics, the "Jumbo" (sometimes called the "Pregnant Guppy") was completed and test flown by Ken Healy. The airlifting began. A company was formed, called, as I remember, Aero Spacelines. It performed this essential service for several years.

CHAPTER 12

Braniff International and the Vietnam Jet Airlift

> *I sign on as Navigator for Braniff International Airways just as the Vietnam war gets serious.*

As the Vietnam war heated up, the U.S. government needed a way to get troops from Travis Air Force base in northern California to Vietnam. In 1966, Braniff International Airways decided that their six new Boeing 707-320 jet airplanes could be put to more profitable use on the Military Airlift Command (MAC).

Until that time, Braniff had been a conservative regional airline, basically serving the central portion of the U.S. and South America. In 1965, a new management under Harding Lawrence expanded the airline to an all-jet fleet, broadcasting the change with a great advertising campaign, multicolored planes, and outlandish flight attendant uniforms. Almost overnight, Braniff became the most talked about airline in the country.

Braniff had never flown the Pacific, and their knowledge of navigation over water for long distances was limited. The

707-320 jets were equipped with a new Doppler radar navigation system which could be updated by LORAN, the electronic positioning system, and operated—theoretically—by the pilots. Again, FAA regulations required that each plane carry either two Doppler-qualified crew members or a licensed navigator.

Captain Joe Dean, Chief Pilot of this new MAC operation, obtained the services of about twenty-five professional navigators contracted through International Air Services Corporation (IASCO). This company had been supplying crews to Japan Air Lines, Lufthansa, and other smaller national airlines that were starting their own international airlines, but did not have trained personnel of their own. World War II had cost Germany and Japan many of their experienced people, and they needed help. IASCO supplied these airlines needed personnel who stayed on the IASCO payroll. Then, as the IASCO clients trained their own people, the contracted personnel could be terminated.

Although I had considered myself retired from flying, the thrill of once again becoming a part of this new jet age operation was compelling. I signed on.

Flying Made Easy

The high-altitude Boeing jets flew above cloud cover. That made the stars more easily available for positioning. The cabin was pressurized, so navigator's no longer had to expose themselves to a rush of frigid outside air while they made celestial observations. Periscopic octants made for a more comfortable environment. All celestial sights now had to be precomputed so that the stars could be found in the limited field of view. The days of flying sixteen to twenty-two hours were over, and we never referred to them as "the good old days."

No Errors

Airline crews are shuffled every month, a good way to guard against complacency. As newer equipment gets more complicated, pilots cannot be expected to make in-flight repairs. Redundancy is the answer. For example, if you put three Inertial

Navigation Systems (INS) aboard, and all three agree, then the pilot can assume that all three are correct. If one malfunctions and the other two agree, it is still reasonable to believe the remaining two are reliable.

Computers do what they are instructed to do by the crew. In this case, the instructions should be redundant. They should be loaded by one pilot reading from the flight plan, then by another pilot entering the data as read to him, while a third pilot double checks them both. This will help eliminate human error.

When crew members get complacent and trust someone else to do something, they are asking for trouble. A Korean 747 was shot down over the east coast of Russia for straying off course. I have always felt that it was "finger error"—that the wrong data was entered into the navigation system and no one double checked it carefully. In aviation, errors must be quickly corrected even if it means a loss of face. A small price.

Essential Again

We were sent to Dallas for a two week refresher and instructed in the new Doppler (partially computerized) navigation system. In theory, we were to be instructors only until every crew was duly qualified in Doppler. That meant the job was probably not going to last.

But I was in for a happy surprise. In the airline business, job vacancies are bid for and awarded according to seniority. There is a constant shifting of personnel as pilots bid for routes that would upgrade their status or that they considered more desirable. Even trip assignments were bid according to seniority. This prevented any certainty that two Doppler-trained crew members would be on board so the plane could fly without a navigator. The net result was that, even after two years, navigators were still used on this airlift.

The vice president of flying operations was General Herman Rumsey, the same Colonel Rumsey I had been sent to Tokyo to meet prior to the Overseas National 1959 Tokyo Military Airlift.

I had become quite an admirer of the Chief Pilot, Captain Joe Dean, an ex-Marine Colonel, so I moved into an adjacent office and offered my services to monitor the pilots' navigation performance. The word got out that I was checking all their charts which made them redouble their efforts and not cheat. I actually couldn't possibly have checked them all, but I did do a good sampling, and it kept everyone on their toes.

Eventually, my work got me on the Braniff payroll with privileges not enjoyed by contract personnel.

Serving Vietnam

The MAC operation from Travis flew via Honolulu, Wake Island, Guam, and Clark Air Force Base in the Philippines to various bases in Vietnam. These large jets were never left at a base in Vietnam longer than the time it took to unload, refuel, and reload. In the three years of the airlift, we never even had a meal in Vietnam.

The bases we served were at Saigon, Ben Hoa, Phu Cat, Cam Ranh Bay, and Danang. Some return flights went via Okinawa and Japan, others by way of Clark, Wake, and Honolulu.

Whenever we stopped in Tokyo, I was impressed by its recovery from the war and how many improvements had been done since my last time there. I always tried to see the Imamuras and found their personal progress heartwarming.

Our passengers out of Vietnam were often soldiers being relieved from duty for R & R (rest and recreation). Some were going home to the states to their families. Our crew rest stop in Okinawa was at a small hotel managed by a middle-aged Japanese who always got out of bed to greet us (in his underwear). We would have a couple of beers which helped us sleep better after a long flight.

Near Naha on Okinawa, we saw large caves with stalactite and stalagmite lime formations, much like those at Carlsbad Caverns. During World War II, hundreds of Okinawans found refuge from the war by hiding in these large, moist caves. While touring this area, we noticed a great deal of activity about 150 yards off shore. Men in small boats were clubbing several dozen struggling porpoises that had become entangled in their fishnets. It was not pleasant to watch. Each tiny boat was nearly filled by one porpoise. I wondered if this was a routine event or an unusual windfall for the villagers.

On one trip to Okinawa, a very young soldier came into the cockpit after we had taken off from Saigon. He was a very wholesome, midwestern farm boy, just relieved from a grueling tour of duty, and he was dazzled by the aircraft's instruments. We learned that he had no family to visit states-side and no destination. When we deplaned at Kadena Air Base for a crew change and overnight rest, he came along although the plane continued on to Tokyo. We found out later that he had spent his entire R&R time at that small hotel. He had no one at home who was expecting or missing him.

Many of these young fellows probably didn't really know where "Nam" or Okinawa were—which in no way takes away from their loyalties and trust.

"Oh, Thank Goodness!"

On our approach to a landing in Vietnam, a flight attendant remarked that she saw flashes of lightning below and was very concerned. She was told, "Honey, that wasn't lightning. They were just shooting at us." Paradoxically, she was greatly relieved, not realizing that tracer bullets can have quite a sting.

What Color Is Your Airplane?

Braniff commissioned the artist Alexander Calder, famous for his mobiles, to decorate several of their planes. The designs were typical Calder, a bit "far out" and controversial, but that was the objective. These "jelly bean" colored airplanes were unusual to say the least. Planes were painted brilliant yellow, green, blue, and red. When any Braniff "jelly bean" airplane

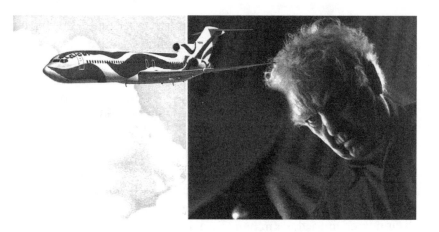

Braniff's Boeing 727 "Jelly Bean Plane"
designed by Alexander Calder, 1973

flew abeam of a Coast Guard ship between California and
Hawaii, it became customary for a female cabin attendant to
greet the sailors on Coast Guard vessels at sea by radio with a
few minutes of light talk.

One of the first questions the sailors always asked was,
"What color is your airplane?" Often they had a ship's pool to
guess the color of the next plane.

In later years, other airlines decorated their planes, but
Braniff's jelly bean planes were the first to deviate from the
conservative formality intended to convey an atmosphere of
military discipline.

The Calder-decorated planes generated crowds at all our
South American stops. People came to see the huge flying
murals. It was a fun time. Cuisine and in-flight service on
Braniff's South American routes were superb. Their ticket prices
were substantial, and their passengers were used to the best.
Braniff did not disappoint.

Wild, Wild West

The volume of crews laying over at Clark Air Base in the
Philippines was soon greater than could be accommodated by

the military. Angeles City, just off base, was more than a bit scruffy, lacking first class facilities—or even third class ones. Just outside the main entrance to Clark, a series of small hotels had been built by Filipino entrepreneurs. Airline crews, lacking anything better, billeted there, but it was not the most hospitable location. One pilot had been robbed, stripped of his clothes, and left in a field a few miles away.

One morning in 1967, while I was having breakfast at the Hotel Plaridel with our local Braniff manager, two armed men with drawn guns threw open the door. It was just like a scene in a Hollywood western. The men glanced around, then disappeared into the kitchen. In a few moments they returned, marching the cook ahead of them at gunpoint. My friend and I tried to be very inconspicuous throughout the incident. We never found out what happened to the cook.

Guam

Guam was a frequent stop, with occasional layovers for crew rest. By the mid 1960s, a number of luxury hotels had been built on the beach at Agana. In stark contrast, some Japanese soldiers were apparently still hiding in the cliffs near Agana, unaware that the war was over—or perhaps too embarrassed by its outcome.

As the closest tropical island to Japan, Guam had become a popular wedding site for hundreds of Japanese tourists and celebrants eager to take tours and photographs. Near one hotel, World War II military equipment had been reassembled from scrapped gear, and smiling tourists posed for pictures in front of this counterfeit combat area.

The nearby islands of Saipan and Tinian were under Japanese control before the war, when the Pan American Clippers made their first stops at Guam. The Japanese protested when deviations from course took the Clippers over these islands. Ted Hrutky, a radio operator on those early trips, recalled that "inadvertent" flights over these islands revealed Japanese fortifications that shouldn't have been there.

Ross Perot's Christmas Dinners

Ross Perot, the Dallas millionaire, wanted to deliver Christmas dinners to prisoners of war in Hanoi, Vietnam. I planned to navigate the trip, but it never happened. The Viet Cong wouldn't give Perot permission for the mission.

Hi, Honey, I'm Home

Inbound from Honolulu to Travis Air Force Base, about twenty five miles out, Captain Cecil always made a 360-degree turn at several thousand feet altitude near Napa where he lived. This alerted his wife that he was back from Vietnam so she should come pick him up. He saved an hour getting back home, but it cost the airline about $50 in extra flight time.

CHAPTER 13

South of the Border

> Hijacking, revolution, and other
> adventures in the Caribbean and
> South America.

Braniff was starting to look like a permanent source of employment for me. In December of 1969, as the MAC Viet Nam airlift was tapering off, Braniff transferred me to their main operating base at Love Field in Dallas, Texas.

Hijacked

No one had ever heard of hijacking an airplane. Then suddenly, like hula hoops and Tickle-Me Elmos, they became an international fad—but with more lethal results. During 1969 and 1970, so many planes were hijacked that the FAA put Federal Marshals in plain clothes on the most vulnerable flights.

A telephone alert came into Braniff Operations Control Center in Dallas in early 1970. A DC-8 flying from Mexico City to San Antonio, Texas was being hijacked. A passenger had handed a note to a flight attendant who passed it on to the pilot. The note said:

This is a hijacking. Do not continue your approach to
San Antonio. Go back to Mexico. We want $100,000
in unmarked U.S. currency. Then fly us to Algeria.

The flight from Mexico City did not require a navigator, so none was aboard although the aircraft had the necessary navigation equipment and over-water gear that could take it to Algeria. It was immediately decided that a management crew would be formed to rendezvous with the hijacked airplane wherever it landed, relieve the crew, and then play it by ear. I was in the Chief Navigators office and available, so there was no question about who was going to be the navigator. We figured we would probably have to go Algeria via South America, crossing the Atlantic from Rio De Janeiro, Brazil.

While we tensely awaited further word, the DC-8 turned around and landed in Monterey, Mexico. The $100,000 was quickly generated by a Mexican bank and handed over to the young hijacker in a paper sack. The aircraft was fueled for a flight to Lima, Peru and departed with its original crew and all its passengers still on board.

That evening, the rest of the management crew and I took the regular Dallas-Miami flight, planning to connect with a scheduled Braniff flight to Lima. Harry Pizer, the airlines' Chief of Security, got a navigator's uniform so he would appear to be part of the relief crew. Harry, a seasoned cop from an eastern city, was armed and ready to get rough if it was necessary. I think he hoped it would be.

The FBI quickly learned the hijacker's name and that of his female Mexican companion. They researched his background, found he had served in the Navy, and then contacted his mother.

There was a psychiatrist in Dallas who had interviewed most of the hijackers already jailed in the United States and written a book about them. He was brought in as a consultant. Telephone lines were held open between Braniff headquarters, operations offices in South America, the psychiatrist, and the FBI.

It was quickly discovered that the hijacker had recruited his female companion at a party in Mexico City the night before. It certainly didn't look like a well thought-out plan. We realized we were probably not dealing with religious zealots or political terrorists, people willing to die for a cause. However, they might be on drugs and must still be taken seriously.

As our plane got closer to Lima, we started asking ourselves a jumble of questions. What are we getting ourselves into? Does he really have a bomb? We knew the plane had a fire axe in the cockpit. Should we try to use it? Would the passengers be held hostage? Would we get some rest before going to Algeria? How would we react under extreme pressure? Things were moving pretty fast, but we knew we'd find the answers to our questions soon enough.

As the chase progressed, people in the various airline departments eagerly offered suggestions for bringing the hijackers down. The catering people urged that drugged food and drink should be brought aboard. Engineering types suggested maintenance malfunction or deflating the tires. Security people would not be reluctant to order a shoot out.

However, the psychiatrist suggested doing nothing heroic or cute. He would advise us in due time.

When we got to Lima, we found that the hijacked aircraft had already landed, refueled, and departed for Rio De Janeiro. Varig, the Brazilian national airline, had an immediate departure for Rio and took us along as their guests. Braniff had never flown from Rio De Janeiro to Algeria, so they didn't have the navigation charts I would need to cross the Atlantic. While we were en route to Rio, Varig supplied us with these charts and advised us on the best routes.

The hijacked plane started to land at Rio De Janeiro, but the presence of Brazilian military surrounding the runway frightened the hijacker. He demanded an immediate pull up. Brazilian fire trucks were coming down the runway as the plane tried to become airborne. The pilot was able to miss the oncoming trucks, and the plane headed for Buenos Aires in

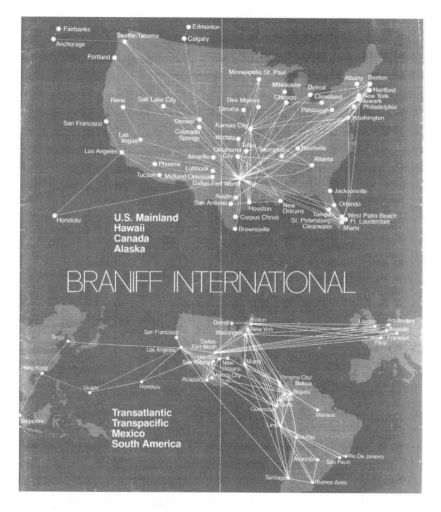

Braniff routes in North and South America.

Argentina without refueling. In the haste to become airborne, cockpit procedures were not followed, and the windshield cracked. This wasn't life-threatening, but it certainly wasn't good.

As soon as we got to Rio, we found that we now needed to go on to Buenos Aires. However, unlike Lima, Buenos Aries had no immediate scheduled departure for our next destination. We chartered a small Brazilian Air Force jet to take us. We were thoroughly exhausted by this time and faced the prospect of a continuing series of flights under very trying circumstances.

We finally caught up with the hijacked plane at Buenos Aires. It was late in the evening. The blue DC-8 was parked in an area away from the terminal. The hijacker had gained enough confidence in the flight engineer to let him leave the plane to relay his demands as well as requests for specific food items. The telephone lines between Braniff headquarters, operations in South America, the psychiatrist, and the FBI were still open, and there was a constant dialogue about what should be done next. Every one had some suggestion for foiling the culprits.

The Dallas psychiatrist insisted that drugs and alcohol were a factor, and that the hijackers would not be able to continue beyond sunrise of the third day. It was decided to exchange crews on the morning of the third day and fly to Algeria. We would be introduced to the hijacker as the "fresh crew." In the meantime, we went to a hotel for a few hours' much-needed sleep.

Then, exactly as the psychiatrist had predicted, the hijackers surrendered at sunrise of the third day. Their threatening package was not a real bomb. Both hijackers were given prison terms in Argentina. The Mexican woman accomplice later married her Argentine defense lawyer. After the man got out of jail, he was retried in Mexico.

Instead of flying to Algeria, both Braniff crews brought the plane back to Dallas. There were no spare windshields in South America to replace the cracked one, so we had to fly home that way. On the flight back, the $100,000 sat on the navigator's table in a plain paper sack. Dallas television and newspapers had given the event considerable coverage, but no one met us when we arrived at Love Field on a Sunday morning. The hangar was empty. There was no one to relieve us of the paper sack. Harry Pizer, the Security Chief who had been with us throughout the ordeal, agreed to deliver it to the company the next day.

Several weeks later, we were invited to attend a presentation by the psychiatrist who had so accurately forecast the events. Braniff's training department produced an instructional film for all flight personnel on the best way to handle hijacking threats. This film was distributed free to other airlines. Some weeks later, an Alaska Airlines flight attendant just out of training boarded her first flight. It was hijacked. She followed the training film instructions, saying things like, "This is going to upset my mother!" She talked the hijacker into surrendering. A gutsy girl just doing her job.

The Mysterious Passenger

In the summer of 1968, my wife, Ellen, and our daughter Dulce, then seven years old, took a Caribbean cruise on the French Lines' Antilles. We flew from San Francisco to San Juan, Puerto Rico where we visited Ellen's father, Ben, now retired. The cruise started in San Juan and was scheduled to make stops at St. Thomas; the U.S. Virgin Islands; Guadeloupe; Martinique, Caracas; Venezuela; Curaçao; and back to San Juan. These cruises sail at night and reach the next port by morning which leaves the whole day for sightseeing and shopping.

In the dining salon, we noticed a tall, very distinguished Black man at an adjacent table. He was always alone and, despite the tropical climate, wore a navy blue suit and tie. The night before we were due at Fort du France in Martinique, our waiter approached and explained in broken English—this was a French-speaking ship—that this gentleman had been admiring our small daughter. With our permission, he would like to present her with a doll from his island of Martinique, where he was disembarking the next day.

Dulce had a small plastic puzzle at the table. Guessing that the man might have children, she sent it to him, and we replied that Dulce would be pleased to accept his doll. The waiter, who by this time needed roller skates, returned immediately with the gentleman's offer to take us sightseeing the next day in Martinique.

At the agreed time, he met us at the foot of the dock with a late model car, a bag of oranges, the promised doll, and a teenaged girl who spoke some English. We drove around the island on very narrow roads. The speedometer was in kilometers which gave Ellen the impression that we were going twice as fast as we actually were. At each village, we'd stop and our host would disappear for ten minutes, return, and drive us on to the next town.

Our host was obviously well educated, and we guessed he might be in some kind of construction business. After our tour, he took us to his home for refreshments. On the wall was a photo of a very lovely French woman. They had met while he was a student in Paris, married, and returned to Martinique, but apparently something happened and she returned to France. He now had a live-in girlfriend who had borne him two boys about eight and nine years of age. One of the boys proudly showed us a picture he had drawn of his impression of the United States: the Grand Canyon surrounded by skyscrapers and cowboys. On the mantle was a stuffed dog. Not a toy, but literally a stuffed dog. It had been the family pet.

Our host drove us back to the ship after another tour of Fort du France. It had been a very pleasant afternoon.

At breakfast the next morning, as we sailed toward Caracas, the waiter asked us how we had enjoyed our day on the island. "It was great," we said, and thanked him for facilitating it. He then told us that he had been a bit worried about us. "France is a small country, and they needed their departments." (We would call them "colonies.") It turned out that our host was the revolutionary leader agitating for Martinique's independence from France. If the island became autonomous, he could become its prime minister. In the meantime, he made this cruise every year, always first class and always alone.

Like the Back of Their Hand

That same day, we were invited to visit the ship's bridge. Naturally, I was eager to look at the navigation chart and to compare their methods to those I used. But there was no chart in sight. Perhaps the Antilles had made the trip so often that one wasn't necessary. It looked as if they were dead-reckoning, using their compasses with corrections for water movements. I remember remarking to my wife that the navigation was quite casual.

Almost a year later, we noticed a full page picture of the Antilles in Life magazine. It had gone aground on a reef in roughly the same area we had been, a total loss.

When the Rich Go Barefoot

The next port after Caracas was Curaçao, a small Dutch Island where Venezuelan oil was being refined. The buildings are all soft pastels, unique and charming. Years ago, a local governor had decreed the color scheme, believing it easier on the eyes. It was certainly a tourist attraction.

The roadway between the ship and the commercial area of Curaçao crossed a pontoon bridge. When vessels needed to pass through the shipping channel, one end of the bridge was towed aside. There was a bridge toll of a few copper coins—but only for those affluent enough to wear shoes. But it seems that every society finds loop holes to beat the regulations. Those who could afford shoes simply took them off and crossed the bridge, footwear in hand.

Peruvian Know-how

A Braniff captain, who had spent years flying in South America, acquired some land in Peru that he planned to level and develop into an irrigated agricultural project. It would be an expensive undertaking, requiring imported American-made earth moving equipment. A Peruvian worker learned of the plan, and informed his boss that it wasn't necessary to go to that expense. When the pilot returned to his property, the land was leveled. No machinery had been used, no expense incurred, and all was in order. The Peruvian had used ancient methods of controlled erosion from a nearby stream.

A Roving Journalist

We had a young copilot named John Nance who flew the South American flights. Likable, thoroughly responsible, and well thought of by his peers, he did not fit the mold of the typical airline pilot. Many pilots used layovers for personal business activities. John was a roving journalist, carrying a portable typewriter instead of a briefcase. Whenever there was an earthquake, uprising, or newsworthy event, John would call a news network in the states and give an on-the-spot report.

After Braniff's demise, John wrote a fascinating book, *Splash of Colors*, about its rise and fall. Throughout the 1990s, he continued flying and reporting. It was not unusual to see his televised comments after some aviation incident. In 1996, his novels like *Pandora's Clock* have been featured on numerous television specials. He was factual, honest in his reporting, and a credit to the profession.

No Charts for L.A. to Lima

Braniff had been servicing South America, flying over land. Flights from Los Angeles to Lima, Peru operated via Mexico City, Guatemala, Panama, and Quito. But the DC-8's were all equipped with over-water gear and a navigation station. A direct flight over water could save twenty-three minutes and eliminate the expenses of communicating with the Central American countries flown over.

We got permission to fly direct, but soon found out that there were no published aeronautical charts for the route. There had never been any demand. The LORAN coverage for that area had to be approximated, and the results were rather poor. I drew a chart, hopefully extending the LORAN lines into the uncharted area, ordered blueprint duplicates, and got permission to navigate a direct Braniff flight from Los Angeles to Peru to see if the results merited a change in their over-land policy. They did.

Our flight route took us abeam of the Galapagos Islands on the equator. Our infrequent flights did not warrant the expense of expensive state-of-the-art equipment, and, without reliable LORAN and the aids available on more traveled routes, we had to revert to celestial navigation. It was like the early Clipper flights—pre-World-War-II methods applied to jet equipment. And it worked. We continued the experiment with a new system called Omega, which had not yet been approved in 1979. Our Los-Angeles-to-Lima direct flights paved the way for Los Angeles to Santiago, Chile by ocean routing, which saved even more time and money.

Years later, I found that the airline was still using my photocopies.

Such Different Lives

A fair young man waiting to board a flight from San Francisco to Lima really stuck out among the predominantly South American passengers. He told me he had grown up in a California town near Turlock, and was back in the States visiting his mother. He had married a Bolivian Indian and now lived in La Paz where he worked in a small shop that produced springs for cars.

Several months later in Lima, I noticed this same young man changing planes. His mother had died, he said, and he had been back to help settle her affairs. When we landed at El Alto Airport in La Paz—the world's highest airport at 13304 feet—I watched as his Indian family rushed to welcome him. He looked completely at home. Here was someone so similar to me who had chosen such a different life and completely adopted another culture.

Blood and Gold

There is a small museum in Lima, Peru which contains an incredible number of golden Incan artifacts. Much of the plundered gold of these areas was taken back to Spain and melted down, but here some of it remains in the form of ceremonial jewelry of great beauty. Large panels of pure gold are displayed beside the very weapons that led to their loss.

Mark Farmer at Braniff, the New York-Rio run.

In a nearby glass case, there is a human skull with a set of quartz false teeth, individually fitted into place. Next to diamond, quartz is the hardest mineral known. What secrets, patience, and craftsmanship these ancients had!

Songs of Insurrection

As we approached a large South American city in the mid 1970s, the cabin attendants invited me to join them at a concert that evening. The name of the artist was not familiar to me, but seemed to mean a great deal to them. I accepted, and they picked me up at my hotel.

The theater was jam packed with about a thousand celebrants, predominantly younger people. I didn't speak the language well enough to understand the lyrics, but each song was greeted with tumultuous enthusiasm.

As the evening progressed, I realized that some message was being shared by the audience. This was a country with stern restrictions on what we consider "free speech." It was illegal for anyone to gather with the intent of expressing political views that might be considered critical of their unpopular government. However, despite the risk, this artist had found a way to express these views in a way that might escape the authorities. I was very impressed. These young folks seemed to enjoy the vicarious thrill of outwitting the political police.

Unexpected Encounters

On one trip over the Amazon, I looked at the passenger manifest for a charter by an agricultural organization. To my amazement, I discovered that many were from my home town, Turlock, people I'd known years before. They had prospered, and I was proud of their success. It was a surprise to run into them thousands of feet above the Brazilian jungles. But I had an even more surprising encounter with someone from the past.

Carnival time in Rio de Janeiro engulfs the city in a joyous bedlam, exceeded only by the frenzy when Brazil contends for a world soccer title. We arrived at Rio on a special charter during the height of Carnival celebrations. The Braniff manager met us with apologies that, because we were not a regularly scheduled flight, there were no suitable hotel rooms to be had. However, he had finally located space in a shabby, run-down hotel in the dock area: "Just for one night," he said, "No air conditioning, no eating facilities. Be careful. You'll have to fight your way past the prostitutes to go to a nearby restaurant. Sorry, but it's the best we could do."

His description was completely accurate. The next day, as we were checking out, I turned to greet our pilot who was standing next to a middle aged bellman. This bellman was short, fat and bald. He asked where we were going and we said New York. He replied, "Have a good trip." The Captain, eyeing the man's shiny head, teased him by saying, "Where's your barber?" This seemed a strange comment, but the bellman answered jovially, "My barber is in Egypt."

I could hardly believe what I was hearing. I asked him if he was the same man I'd met in the restroom of Shepheard's Hotel in Cairo in 1946. He'd handed me a towel and said, "Where are you going, sir?" I said, "New York," and he'd replied, "Have a good trip."

"Are you the same man?"

"Yes, Yes" he said, and excitedly ran out to hail several cabs to take us to the airport. I suspect that, after Shepheard's Hotel had been burned, he had parlayed "Where are you going?" and "Have a good trip" in many languages, into a career that took him to many lands. Twenty-seven years later, our paths had crossed again.

You Can't Afford It

The cabin attendants for South American flights were from the various countries being served. This was by government agreement, and it made for better service linguistically for Braniff's mostly Spanish- and Portuguese-speaking passengers. Many of the hostesses were from prominent families, while male flight attendants were often restaurant owners in Brazil and Argentina. Their wages were very low, but their perks and privileges compensated, making it a good job.

We asked one purser, "Where is a good place to eat in Buenos Aires"? He recommended two places, but said we probably could not afford to eat there. He owned them.

Reprisals

One Captain had a reputation for being very difficult with flight attendants. He was thoroughly disliked. After he retired, he learned that, for all those years, they had been spitting in his coffee.

Sitting on Sausages

In the 1970s, inflation was out of control in Argentina. Material goods—commodities—increased in price as money inflated, but a wage earner's paycheck could lose much of its value in a few hours. Early in the mornings, as we drove from the International Airport into downtown Buenos Aires, we saw lines of people waiting outside banks to cash their checks. They were eager to convert the cash into appropriate goods as quickly as possible.

One of our favorite restaurants, La Mosca Blanca, showed the effects of inflation. Every square foot of available floor space was stacked high with food items. The owners kept their entire liquid assets in the form of canned goods, hams, sausages, and other commodities they'd need for their business. I don't blame them. However, it ultimately made for higher prices.

Braniff Airways could not collect money from the sale of northbound tickets out of the country quickly enough to keep up with the inflation. In the late 1970s, Braniff introduced beautiful leather-covered seats on their planes. This might seem like a tremendous extravagance, but it was actually a result of Argentine inflation. Argentine currency couldn't be converted quickly to U.S. dollars, but it could be spent at any time for hides which could be used for seats. Even the airline was forced to play the commodities game.

Improvised Autos

Santiago, Chile in the mid 1970s was threadbare, tired and worn. The climate and landscape were so similar to California, but the cars on the street made you feel as if you'd stepped back into another era. That these ancient vehicles could run at all was a tribute to the improvisational ingenuity of local mechanics. A Braniff employee in Santiago bought a car. When it started running a bit rough, he discovered that one piston was made of wood!

Americus

The international airport for Sao Paulo, Brazil is near the village of Campinas, about fifty kilometers from Sao Paulo. I visited a shop on the edge of Campinas and watched geodes being sawed into sections, largely for export. The proprietor spoke English with a peculiar accent. I complimented his skill and asked where he had learned the language. He said he was from Americus, a number of miles away.

One of our flight engineers, the son of missionaries in Americus, told me about its unique history. Just prior to the American Civil War, some southern plantation owners disposed of their properties, chartered ships, and moved their families and slaves to Brazil. They called their new community Americus. Although slavery is now long gone and the residents are Brazilian citizens, they still speak English as it was spoken 150 years ago. Happily, they have prospered in this new land.

Two Wrists

I always wore two watches. One was kept on Greenwich time (for celestial navigation), and the other was adjusted for the local time wherever we were. In the cockpit one day, an Argentine hostess asked why I wore two time pieces. When I told her, she remarked, "I didn't know they made watches like that."

Complex Problems, Simple Solutions

When you fly from La Paz, Bolivia to Rio De Janeiro, you get a good feel for the immensity of the Amazon river. About thirty-five percent of all the fresh water in the world goes down this river. During flood season, fish can swim across the usually dry land and end up in a completely different river.

This area is the world's "oxygen factory." To prevent its destruction, its care should be left to the indigenous people who already live there. It's their homeland. They know what to do without requiring massive outside expenditures. This would cost practically nothing to implement. I've always been interested in how so-called "primitive" people have really mastered their environments. These native peoples have adapted to their environments without destroying them. They preserve natural resources for their own use, benefiting all of us. In a global catastrophe, these people would certainly survive.

The natural resources of South America equal or surpass those of North America. I suppose there are many reasons why their development there has been so different. They have many small countries, a different concept of democracy, a smaller middle class to provide potential entrepreneurs, and political instability. Brazil has vast resources and is capable of planning and carrying out large engineering projects. But the contrast between side-by-side abject poverty and comfortable affluence is enormous. On one side of the street in Rio, people can be starving, while on the other they are dining on fantastic feijoada (beef, pork, and black bean stew), heart of palm salad, and chop (beer). It was hard to sit in a nice restaurant enjoying ourselves while looking out on such poverty.

South America first attracted foreigners who went primarily to acquire riches and take them back to Europe. North America first attracted foreigners who wanted to become permanent settlers. Is this what made the difference?

CHAPTER 14

Adventures in Africa and Asia

Observations and incidents in Africa, Asia, Europe, and even the U.S.

In January of 1971, Braniff took delivery in Seattle of the hundredth Boeing 747. This airplane, with a capacity of 300 passengers, was to fly a daily round-trip from Dallas to Honolulu. Three inertial navigation systems were installed, which meant that no navigator would be needed. However, the system could not be approved until the plane had made a number of flights. So, although there was no navigation station on the 747 and no way to do conventional navigation, the FAA said a navigator had to be carried until system approval was confirmed. I went to Seattle for the delivery of the aircraft, and rode back on the top deck with nothing to do. However, it made the trip legal.

On the inaugural Dallas-to-Honolulu flight, the passenger list was ablaze with celebrities: governors, mayors, the board of directors—and me with my briefcase in case the system failed. For the return flight, my place was taken by Hal Clements, another person qualified in inertial navigation systems. Because two people on a 747 had to be qualified in INS, either Hal or I would go along as the second qualified person whenever a new pilot was assigned. We'd instruct the new pilot and "check him out." This meant I made many Dallas-to-London non stop polar flights between 1975 and 1980, just riding the jump seat and trying to be helpful.

Tea and Temperament

Airlines are very sensitive to delays. Gate positions at airports are tightly scheduled, and passengers have connecting flights to be met. A delay on a 747 can inconvenience hundreds of people. The most inexcusable delay would be a crew arriving late at the airport without sufficient time to perform pre-flight duties.

On one Braniff 747 flight I was on from London's Heathrow to Dallas, the crew was to be picked up at their downtown layover hotel by a contracted bus. The driver arrived late. Maybe it was his fault. Maybe it wasn't.

As we scrambled on the bus, someone made a sharp comment to him, saying that we urgently needed to be at the airport so he shouldn't worry about the speed limit. I could see that the driver was unhappy.

Halfway to Heathrow, he turned off onto a rural road, parked the bus, and sauntered slowly to a cafe for his "tea break." We sat fuming until his even more leisurely return. Although a tea break was probably covered by his work rules, no American driver would punish other professionals by such behavior. His petulance could have inconvenienced and even endangered many people.

Spectacular Sights

Some of our routings went north of Hudson's Bay and over central Greenland, an awesome mass of ice and snow. The eerie curtains of northern lights that escorted us undoubtedly accompanied the men and dogs who first explored these areas.

Boeing 747

*The "Big Pumpkin," the 100th 747 off the assembly
line. We picked it up at Boeing in 1971 and flew it to
Dallas. The next day we put it into service to Hawaii.
This plane flew without incident until long after I
retired in 1981, never missing its daily round trip
between Dallas and Hawaii.*

Round Rainbows

It is not generally known that rainbows seen from high
altitudes are circles projected on the clouds, below with a
shadow of the airplane in the center.

Airborne Baby-sitting

Braniff had a very liberal ticketing policy for employees.
If space was available, they could buy it with a nominal charge
which barely covered the cost of food. Most employees were
able to enjoy the whole system as a perk not enjoyed by the
public. One pilot in Dallas used Braniff as a baby-sitter for his
two children. He'd calculate how much time he and his wife
needed for a social engagement, put their children on a
scheduled Braniff flight—to Chicago, for example—and have
the cabin attendant see that they were on the next flight back to
Dallas. After their social event, the parents would meet the plane
and take their kids home.

But No Gauguins

In February of 1973, Braniff operated a charter for the
military to transport personnel to Antarctica. Our mission was to

take them from the U.S. to Christchurch, New Zealand. From there, they'd continue the trip by military turboprop aircraft. A crew change was scheduled at Papeete, Tahiti, which meant positioning a relief crew there via another airline. French Polynesia was one area I'd never visited, so I eagerly signed up to be part of the crew.

While we waited for the Braniff plane to arrive, we took a van ride to the Gauguin museum several miles from town. We had all heard so much about the French artist, Paul Gauguin and were eager to see his work firsthand. It was a beautiful spot, but the museum didn't have a single painting by the famed artist! Before he died, he burned all his works that remained in Tahiti. We recalled reading this, but the reality was uncomfortably demonstrated.

The Smell of Grass Growing

In the 1920s, the Farmers were one of the few families in Collinsville, Oklahoma who had a storm cellar. It was located several feet from the rear of the house, ready for use at a moment's notice. The cave-like structure was dug into the ground and had an arched concrete roof, covered with a foot of soil which had been sodded to prevent erosion. We could enter through a lid-like door and down some concrete steps into a dank eight- by ten-foot room. At the far end was a ventilation chimney. Near the entrance, on the right, was an oil lamp for emergency lighting and several shelves where mother stored the fruit and vegetables she had preserved in Kerr glass jars. A drain in the floor was probably connected to the sewer in case of flooding. Built-in concrete benches lined periphery of the room for seating. The mound that covered the cellar made a small hill that we liked to play on.

We children were forbidden to go into the cellar unnecessarily. It was the refuge of last resort if a cyclone was getting too close. I was always worried about what would happen if we were inside and a fallen tree blocked the exit door. How would we get out? And where would our neighbors without storm cellars go? Several times we were alerted to get dressed and be ready to evacuate to the cellar as we watched a funnel cloud approach. It would threaten, pause, and then change course. After the violent thunderclaps stopped and the storms passed, the grass always smelled amazingly fresh and new, a happy message that all was now well.

Then we moved to the Central Valley of California where summers were dry and tornadoes never seen. I didn't regret the tornadoes and thunder, but I always missed the odors of burgeoning grass. I never experienced them in California. Perhaps my senses had dulled as I matured, making it difficult to detect what I had remembered.

Thirty five years later, I used my cockpit jump seat privileges at Braniff to make a side trip to Tulsa where I rented a car and drove to Collinsville. My relatives had all died, their properties disposed of, and a new generation was in charge. There was a rundown, "Tobacco Road" feeling in this once-prosperous town where my family name had once meant something. As I cruised the town, the hills didn't seem so steep and houses looked smaller. I felt a sense of guilt that we Farmers had departed for greener pastures, leaving the community to fend for itself. During that Sunday afternoon, there was a brief shower. Then the sun came out, and there it was! That special, exhilarating scent from the grass, exactly like years ago. I had not lost my ability to detect that subtle, unique sweetness. At our old house, the grassy covered cyclone cellar was still in place. There was a feeling of both renewal and closure as I returned to Dallas where another new world was opening up.

Bomb on Board

Our DC-8 had just departed the ramp at O'Hare airport in Chicago when we were told to return immediately. There had been an anonymous phone call that we were carrying a bomb.

Everyone suspected it was a crank call, but we couldn't take any chances. A detailed inspection found no suspicious device was aboard, but we couldn't depart until the proper reports were filled out and filed. In the next hour, the Chicago police, airport officials, and company representatives generated more paperwork than the phantom bomb would have weighed.

The episode cost several thousand dollars in bureaucratic follow up and delays, and we missed connections at our destination. All for someone's idea of a joke.

A Grim Day

On June 5, 1968 we had positioned ourselves in Berlin to crew a charter several days later. We were free to explore Berlin the next day, and we planned to visit the Berlin Wall and go through it at check point Charlie to see East Berlin. World War II rubble still lined the streets more than twenty years later. The city was shabby and dismal. The only things for sale were a few postcards. The pall was deepened by the news that Senator Robert Kennedy had just been assassinated in Los Angeles. A miserable day.

The Algerian Episode

In 1972, an offshore oil pipeline company in Houston had made a number of trips on commercial airlines to Algiers, negotiating for rights to oil which they hoped to refine somewhere on the east coast of the United States. To show their sincerity and affluence, they chartered a BAC-111 from Braniff for a trip to Algiers. The flight included top executives and their bankers.

The aircraft used was Braniff's corporate executive jet, chosen because it had no markings. It had two jet engines and was luxuriously outfitted for private use. The United States did not then have formal diplomatic relations with the Algerians who had just become independent from the French. The fierce struggle had left bitter feelings all around. During the revolution, the pro-independence faction had been labeled "terrorists" and "traitors." Now they were "liberators" and "patriots." (Later, after viewing their efforts, I agreed with the latter terms, feeling that the Algerians were truly trying to make things work.)

Again, I assigned myself to the crew for this interesting junket. To create the impression that we were with the oil company, we didn't fly in uniform. We could not get Algerian visas in the U.S. so the flight was scheduled from Dallas to Gander, Newfoundland, then to Keflavic, Iceland and on to Paris where the Algerian consulate would issue visas and entry permits. A French lawyer boarded our flight to supervise the negotiations. She had successfully defended Algerian "terrorists" and had become a trusted friend of Algeria. Out of Paris, our flight plan showed us as an Algerian plane and crew, not an American one. In Algiers, we were met by top government officials and bypassed Customs. We were housed at a beautiful resort on the Mediterranean which seemed very much underutilized.

Although this trip was primarily an oil negotiation, it was soon apparent that Algeria had other needs that Americans might be able to fulfill. At the end of the Algerian revolution, the French were literally evicted. This included the professionals who virtually ran the country. French farm managers returned to France, and, without agricultural supervision, the workers fled to the cities. Lacking even middle management, the economy was suffering. One plan to entice workers back to the farms was to build them adequate housing. Could American contractors be encouraged to submit plans for the project? It could be a real plum for enterprising contractors.

We wanted to fly back to Dallas via the Azores, but no one would clear an airplane from Algiers to the Azores, or indeed anywhere in the United Kingdom. We couldn't mention we were Americans because we weren't supposed to be there. Intriguingly, even though diplomatic relations didn't formally exist between the U.S. and Algeria, the American Chargé d`Affairs was still present, respected, and very much a presence. Our aircraft flew to Scotland, landed, but was not permitted a gate position. They were not sure who we were and were not taking any chances. Eventually our aircraft was serviced on the edge of the airport, and the flight continued as an American aircraft.

The oil company chartered another trip to Algiers in September of 1972, and requested the same crew. We flew via

Iceland to London, where an overnight rest stop had been planned.

At London's Heathrow, our BAC-111 was parked right next to a Soviet Russian jetliner whose crew was still on board. By custom, they had to remain with their aircraft until relieved by a fresh crew. Several of us approached the plane for a closer look and were immediately invited to come aboard. We were warmly treated as part of the fraternity and plied with souvenirs from their aircraft. To reciprocate, we invited them to see our jet which was much smaller, but more luxuriously appointed. The only gifts we could offer were playing cards, magazines, and cigarettes. The Russian pilots invited us to join them for dinner at their downtown domicile. We were expected to be part of the oil company group that evening, so reluctantly we had to decline the Russian offer. It could have been an evening to remember.

On this trip, we stayed in the St. Georges Hotel in Algiers. The food and drink at the hotel were first class, as were the listed prices. Each meal check was signed for by a member of the group with his room number. We obviously were exceeding the amount of cash we were carrying, and had much intrepidation about checkout day. It was a pleasant surprise to find that we were "guests" of the government. There was no charge at all.

The St. Georges had brass plates on the doors identifying the rooms occupied by Allied Generals during World War II. During our week-long stay at the St. Georges in Algiers, we were constantly accompanied by Mohammed Bill, a trusted hero of the revolution. He had been very effective in hand to hand fighting with many kills during the revolution. After the French were ousted, he was given an expropriated dry cleaning business. At first, we thought he was assigned to watch us, but later we realized that he had been picked to make certain we were made welcome and our needs properly taken care of.

Mohammed Bill was technically illiterate, but spoke a number of languages. I've noticed that people who cannot read or write usually pay close attention to what is being said, and are very accurate in recalling what they have heard.

Accompanying the oil company group was a representative of a building contractor who planned to make a pitch for agricultural housing. His presentation was composed entirely of slides of prestigious upper income homes, complete with green grass and landscaping. This showed his complete misunderstanding of the realities of housing in desert areas. The Algerians wanted simple, concrete-block cluster housing with common bathing facilities. There was a lucrative government contract for anyone who would do some simple research to fill their needs. Americans got the first chance, and they blew it.

We were invited to spend an evening at a castle-like structure overlooking the city of Algiers. It was occupied by a very high-ranking government official. I recall a large mosaic mural on the wall which depicted the Spanish welcoming the Moors to Spain. Dinner, including a whole roasted lamb, was served on a long table. As guests, we were invited to partake first. We knew that, under Arab customs, we should not use our left hands, and I'm sure that we must have appeared to be hesitatingly awkward.

I was impressed with the great efforts of the government to stimulate their economy. Algeria was an oil-rich country, but without adequate supervisory help to get things done. Every time they drilled for much-needed water in the desert, they kept striking oil.

Dakar, Senegal

In June of 1973, Braniff contracted to deliver a charter plane load of young Peace Corps volunteers to Nairobi, Kenya in East Africa. The flight required a crew change in Dakar, Senegal, the western-most point in Africa. This necessitated positioning a "rested" crew in Dakar so the aircraft, a DC-8, could continue without delay. Because there was no daily service to Dakar at that time, we "commercialed" there on another airline and had several days before our aircraft was to arrive.

My most lingering impression was the starched cleanliness and pride of those we met. We rented two cars to drive down the coast toward Gambia, later of Roots fame. The road was very narrow, somewhat paved, but void of commercialization.

Lunchtime found us about fifty kilometers south of Dakar with no prospects of finding a suitable European-style restaurant. Suddenly, someone spotted a Coca Cola sign near a crude building with a corrugated metal roof. As we drove up, we saw that a squealing pig was tethered outside. We had found a "restaurant." Inside, we were seated on wooden benches and a long table on the packed-dirt floor. Realizing that there was no refrigeration and the facilities were crude, what could we order that would be palatable and yet safe for us to eat? We were near a river and the sea, so we decided to order sole with fried potatoes, plus beer to drink. This probably saved the life of the pig and several meandering chickens.

The contents of the enormous, perfectly arranged platter that came out of that crude kitchen were unbelievable. The filet of sole was delicate and wonderfully seasoned, the crispy potatoes melted in our mouths. Our rating: Food—four stars; Decor—rusty.

It seems that in countries where food is dear, people are more caring about seasonings, preparation, and presentation.

Slaves

The island of Goree, located in Dakar harbor, has a grisly past. It was a collection area for captives being held pending transportation to slave markets where they would be auctioned off. The cells are still there to see. You can hardly stand upright in the stuffy, cave-like rooms which seem crowded though they are empty. The survival rate must have been horribly low. The captives were brought in at one end of the facility, and held until they could exit directly to the gangplank of the next ship where conditions were equally appalling. A horrible reminder of man's inhumanity.

Exotic Food

While eating breakfast in a third-world country, we noted that the specialty seemed to be a bread dipped in egg batter, fried, and served with a sweet sauce. To make conversation, I asked the waiter, "What do you call this?" He replied in perfect English, "I don't know what you call it, sir, but we call it French toast." Touché!

Peter Minuit Wuz Robbed

I don't know the current value of Manhattan island. Perhaps the assessor of New York City can give an estimate if we really needed to know. But if the twenty-four dollars that Peter Minuit paid the native inhabitants in 1624 had been invested at six percent and compounded, it would be worth over two trillion dollars today.

The question is, "Did he pay too much?" Some fear that the trade beads the locals got were junk baubles, foisted on them by con men of the time. However, I have a feeling these beads were probably very beautiful Venetian-type glass beads, the ones with a three-dimensional effect.

When I visited a market place in Dakar, I noticed several of these trade beads for sale. I asked if it was possible to get a complete necklace of them for my wife. The merchant thought it was and suggested I return the next day. True to his word, he had exactly what I had hoped for, a beautiful necklace with twenty-five antique trade beads in good condition. Price: twenty-four dollars, the price of Manhattan. I wondered if there was some kind of price control on trade beads.

The moral of the episode seems to be, "Don't keep your money in trade beads—they don't appreciate." (The necklace proved too heavy to wear, so my wife Ellen had it framed. It hangs in our bathroom, so I am reminded of my transaction several times a day.)

Impromptu Safari

We were approaching Nairobi over Lake Victoria when the ground controllers asked if we had ever been to Kenya. Would we like to go on a safari after landing? Our answers were "no" and "yes." We were met by a Land Rover for a tour of the wild animal preserve outside Nairobi.

There we saw the impressive and elegant Masai tribesmen, as well as those wild animals who were willing to put up with our curiosity. The safari was well worth our loss of sleep before we returned to Dakar.

We returned with the U.S. phone numbers of several Peace Corps kids. I called their parents soon after I got home. It was gratifying to be able to tell them that their children had arrived safely and were in good shape.

A Lesson in (Mis)Communication

My wife and I dropped in to visit a neighbor—I'll call him Jack—who was slowly phasing out his appliance distributing business and looking forward to retirement. The telephone rang. We heard Jack say skeptically, "Oh, yeah? And how are you going to pay for it?" He listened, then hung up in disgust.

"What was that all about?" I asked. He replied that the caller said he was a Saudi Arabian prince and wanted to buy 10,000 kitchen units, paid for with letters of credit. "What is a letter of credit?" Jack asked.

I told him they were as good as cash, and the way to do business overseas. Fortunately, the phone rang again. It was the prince, wanting to know if it was possible to charter a 747 to haul a load of cement to Arabia. Jack turned to me quizzically. "Of course," I told him, "but what a terrible waste of money."

After more phone conversations, Jack made a number of visits to Saudi Arabia and ended up with some construction contracts. He monitored the financial end carefully, but not the actual construction.

The buildings to be constructed were houses with concrete-slab floors and cement-block walls. Jack showed the local foreman how to mix concrete with the proper proportions of sand and cement. "Start the blocks at this corner, and keep on going," Jack told him.

When Jack returned, he was astonished. The mortar between the blocks contained bits of tan-colored paper. The workers had thrown the entire bag of cement into the mixer. Then they carefully constructed a wall down one side of the floor slab. But, instead of following the edge of the slab and turning the corner, they had obeyed Jack's instructions verbatim and "kept going." Also, no one knew to put the plumbing and wiring inside the wall during construction, so channels had to be gouged in the finished walls to accommodate them.

Not putting yourself in the other person's shoes can lead to comical but costly communications errors. Of course, the Arabs hadn't done much checking either. My friend was Jewish.

Expanding into the Pacific

In the summer of 1973, Braniff opened a pilot base in San Francisco to facilitate an expansion into the Pacific, and to inaugurate scheduled DC-8 flights to South America from San Francisco and Los Angeles. I got the assignment to manage the pilot base because I had operated a similar base for Overseas National Airways in 1959-62 and because Chief Pilot Captain Joe Dean was then in Miami.

My compensation as manager was considerably less than that of a senior pilot—possibly why I got the assignment. Even with my new position, I could still function as Chief Navigator, go to Dallas monthly for inertial navigation instruction, navigate to South America once a month to keep up proficiency, and check-ride other navigators for the FAA.

At San Francisco, Braniff occupied a three-room suite in United Airlines' Operations headquarters. The largest room was where crews reported in for flights and stored their uniforms and briefcases between flights. Many pilots commuted to San Francisco from as far away as Texas to avoid relocating their young families for what might end up being a short-term job. Frequent moving was becoming a way of life in aviation, what with deregulation of airlines and the subsequent exploding expansion, new equipment, and new routes. Opportunities to upgrade were frequent. In retrospect, those who chose to leave their families in one place fared better.

Sudden Disappearances

Unexpectedly, we experienced a plague of minor thefts. Helen Callison, my administrative assistant, and I started to receive complaints about missing insignia, uniform parts, briefcase components, etc. The building was secure due to United's operations. We were on the second floor in a sensitive area, so we were very puzzled.

I was in my office after business hours when I heard someone in the crew room. There was no flight leaving, so I assumed it must be a commuter pilot checking in early. I went to see.

A very young man wearing a Braniff uniform was going through a pilot's briefcase. He obviously didn't know who I was, and said quickly that he was just in from South America and tired from the long flight. I knew there had been no such arrival, and I personally knew all the pilots. None were teenagers. I summoned Security who wrote down his name and address, then escorted him out.

Several days later, Helen caught the same boy in the same place. This landed him in juvenile hall. We don't know how he got a Braniff uniform and insignia or made his way into this inner sanctum, but his ultimate object was to go aboard planes, sit in the cockpit, and play pilot. Not a safe thing to have happen. Because he was a minor, he was not prosecuted.

400-Year-Old Steak

I took my daughter Dulce to Amsterdam for a week's vacation. She was eager to see the museums and all the folks in wooden shoes. We stayed in a very old hotel near the palace in the downtown area.

For dinner, I ordered a steak and noticed it was identified on the check by a very large number. When the hotel was first built, the waiters were not literate, so the first steak sold was number one. A running tally had been kept in the four hundred years since. Our meal was unique and without duplication.

Besides the museums, Dulce wanted to go to the house where Anne Frank hid from the Nazis during World War II. She was then the same age as Anne Frank and familiar with her diary and the subsequent movie. We spent an hour in the very rooms where Anne and her family lived. It was a good way for our daughter to observe the sufferings and intolerance of wartime.

Hong Kong

Hong Kong is a city completely at ease with high-rise sophistication, yet still retaining Chinese customs and culture. I will always remember looking out at the city from my hotel room early in the morning. Below me, people were criss-crossing a small park, on their way to work, to shop, or just to enjoy the air. Nearly all paused to do some tai chi aerobics before moving on. There didn't seem to be any leader as this ever-changing group performed their silent ballet, a combination of exercise, dance, and neighborly togetherness. Many of the participants were quite elderly, yet their movements were as smooth and synchronized as those of much younger participants. During the twenty minutes I watched them, some would drop out as their needs were met while others arrived to replace them, all without any conversation or apparent plan. This "performance" continued with an ever-changing cast. Each was doing something for himself or herself, not realizing they were creating something beautiful when seen from above.

I was again in Hong Kong because Braniff had acquired a DC-8 from Alitalia Airlines in Italy in April, 1979. That's not quite as illogical as it seams. The Italians, to comply with American standards and make the plane compatible with the Braniff fleet, delivered the DC-8 to Kai Tak airport in Hong Kong for conversion at the local maintenance facility. Kai Tak's expert mechanics were paid ten dollars a day and did first class work. I was one of a management crew that gladly flew to Hong Kong to bring the "new" plane back to Dallas. We arrived several days before the estimated date of completion. Then there was a further week's delay, so we had time for a great vacation.

The completed plane was equipped with Inertial Guidance navigation gear, and the winds looked favorable, so we decided to attempt a nonstop flight to Dallas. If we succeeded, we'd set a long distance record for the DC-8. We flew abeam of Japan, "up" to the Aleutian Islands, and "down" towards Seattle. But, as we neared Seattle, we decided our fuel reserves were insufficient to continue. We stopped to refuel and reached Dallas a few hours later. The Italian plane had been flown three-quarters of the way around the world to take advantage of lower cost Hong Kong labor.

Souls Suspended in Space

The people aft in the cabin were rarely making routine trips, so it was interesting to speculate why they were aboard. In First Class, people were usually more affluent and demanding. In the aft section, there were always some students, some family groups, salespeople, and sometimes refugees seeking a better life. Some might be children who could not yet speak English, ready to assimilate into the American scene and take their places as new citizens. Some would be on holiday, some on family emergencies. And I'm sure we had our share of smugglers and people running from families and even the law. All these people, sealed in a capsule and trusting that the various institutions upon which we depend have provided for their safety and timely delivery to destination. Intimately batched together as a temporary unit, only to be broken up on arrival and never reassembled.

Military passengers were always the best—and the hardest. If we were taking them to their tour of duty, we hoped they'd all be going home with us in good condition. But we knew better.

The Pan American seaplanes (Naval Air Transport Service) of World War II were for top priority personnel, mail, and cargo. Usual loads were eight to twenty five souls. In the Korean conflict DC-4, we carried about seventy-five people per flight, not all high priority officers. The Boeing 707s of the Vietnam war carried 165 passengers. The Desert Storm airlift to Saudi Arabia used Boeing 747s and could carry upwards of 350 troops at four times the speed of our original seaplanes.

CHAPTER 15

Last Flights

> *A forty-year view from the cockpit.*

The San Francisco base was closed in early 1981. I didn't want to relocate to Dallas when I was so near to retirement and with an airline whose days were numbered. Therefore, I chose early retirement at age sixty-three. (Only pilots had to retire at age sixty.)

I had started out as one of the first professional air navigators. Forty years later, I ended as one of the last.

Starting in the 1990s, a small group of Pan American navigators began getting together each year for a reunion. They are a splendid group and we have many memories in common.

What's Different Today?

My forty years in aviation started with the first Pacific aircraft navigated by adapted marine methods and ended with Boeing 747s with inertial guidance. On our early Clipper seaplane flights to Hawaii, our job was to find the destination using the most efficient route. There was little or no other air traffic to worry about. We spent long hours in the crowded quarters of a cockpit capsule, no place for anyone

Pan American navigators' reunion in San Francisco, 1983

claustrophobic, yet, we saw the world from this cramped cubical. Although the work could be demanding, it was always rewarding. Many things happened that enriched my life.

Navigation today is considerably different. On the plus side, computer flight plans and satellite weather data mean there is no problem finding a destination. Great ground support is instantly available at any time a problem occurs, unlike the early days when crews had to rely on their own diagnostic abilities to solve an in-flight abnormality. And, with flight times of five or six hours, crews are more likely to be alert than in the old days of twenty-hour flights.

On the minus side, the skies have become more crowded, and airspace has to be shared. Decisions often must be split-second. No time for pondering the best course of action. When two planes are heading for each other at 500 miles per hour, they have a closure rate of 1000 miles per hour. That leaves little time for correction.

The original concept of international flying was to travel to far away places in less time than ships would require. Today's

B20 THE WALL STREET JOURNAL FRIDAY, SEPTEMBER 12, 1997

American Air Is Found Guilty In Colombia Crash

SCOTT McCARTNEY
Staff Reporter of THE WALL STREET JOURNAL

A federal judge found AMR Corp.'s American Airlines guilty of "willful misconduct" in the 1995 crash of a Boeing 757 jet into a mountain near Cali, Colombia.

The decision by U.S. District Judge Stanley Marcus in Miami lifts the $75,000 international limit on individual damage claims stemming from the crash that killed 159 passengers and crew. The judge granted a motion for summary judgment in the liability case that had been scheduled to go to trial next Tuesday.

In his decision, Judge Marcus said the airline was guilty of willful misconduct and "reckless disregard" for passenger safety because its pilots continued a nighttime descent in mountainous terrain knowing they were off course.

Plaintiffs' attorney Victor Diaz in Miami said the ruling marked the first time a federal court had determined liability without a trial in an aviation case. Plaintiffs can now proceed to individual damage trials, he said.

American said it would appeal. In a statement, the carrier said that "neither the airline nor our flight crew intentionally took action that resulted in the accident. We regret that Judge Marcus's ruling will prevent us from presenting evidence to a jury that proves our pilots were clearly not guilty of willful misconduct."

In his ruling, Judge Marcus noted that American's own training manual warned that most aircraft that hit mountains in Latin America did so because crews apparently did not know where they were. "The undisputed facts, and the many points conceded by American, simply leave no room for a reasonable jury to find that the pilots' deliberate decision to continue to descend the aircraft from an off course position at night, in an environment known for high terrain, . . . was anything other than an act of willful misconduct," the judge said.

American Flight 965 crashed Dec. 20, 1995, after pilots incorrectly programmed the plane's navigational computer while approaching Cali. Investigators said the crew, following printed charts, punched "R" into the computer trying to select a navigational beacon called "Rozo." But the computer offered "Romeo," a beacon behind them near Bogota, Colombia, as the first choice under all the "R" listings. The error turned the jet toward mountains, confusing the cockpit crew, while the descent continued.

Under the international aviation treaty, damage claims are limited to $75,000 a person unless an airline is guilty of willful misconduct. American and its insurer previously said they would waive the $75,000 cap. But while the airline admitted its crew had made a negligent mistake, it contested that the crash resulted from willful misconduct.

Why computers will never totally replace common sense.

jets have figuratively pulled the continents closer together. Just a few decades ago, no one planned a two-week vacation overseas. The airplane has made it practical for the middle class to vacation or study anywhere in the world. Combining low coach fares and the moderately priced hotels such as those erected by entrepreneur Roy Kelley of Honolulu, millions of ordinary people can enjoy what used to be reserved for millionaires. The same is true for people in Europe and Asia.

Logic in the Age of Computers

When pilots were introduced to the wonders of computerized inertial navigation, a full school day was spent in reviewing the Polynesian methods and China Clipper procedures—simple fundamentals which could be lost if people enter unchallenged and possibly erroneous data in a computer and then accept the results without question. What if something happens to the computers? The crew should be able to fall back on traditional methods.

My father could add up his store's daily receipts with uncanny speed and accuracy in his head, without an adding machine. I am from the slide rule generation, used to an adding machine or calculator. I didn't acquire his mental skill because there was no need to, but I did have to use logic. The drawback of using a slide rule or navigation circular calculator is the lack of a decimal point. The operator must decide logically where to put it—whether the answer is 6,500 or .065. I hope today's computers never stop people from challenging the logic of answers.

You Guys Know Who You Are

Navigators from World War II, Korea, Viet Nam, United Nations Mercy Airlifts, and Desert Storm will find something of themselves and their lives in this book. That we were able to survive long missions under primitive operating conditions for so many years with so few incidents is a tribute to Pratt and Whitney's engines, Boeing, Douglas, Martin, and Lockheed.

Hundreds of you guys have stories similar to mine. I hope you'll smile as you read this and say, "Yeah, I know just what he's talking about." We may have flown at different times and in different places, but you've been there too.

Some were hired for an emergency, flew, and then dropped out. They made themselves available during the crisis and then returned to their civilian pursuits. From time to time, they came back when they were needed—or when they needed some extra cash. One navigator I know went back to flying long enough to earn the cost of a tow truck for his auto repair business.

Some had become ranchers and farmers who were called away at harvest time. There were court reporters, vending machine operators, office supply store owners, salesmen, teachers, and orchardists who came when they were called. The long, routine hours were broken only occasionally by those challenging moments that maybe only the navigator or flight engineer were aware of. All these men would have stories to tell, though rarely the dramatic kind that Hollywood prefers.

20/20 Hindsight

The older person who looks back knows how every incident ended. There is no risk. Everything came out all right or it didn't. Survival itself is a measure for some reflection. All chapters have had their endings, so it's possible to be secure and confident.

Not all the principles I learned in childhood proved accurate, but, on balance, they offered good guidelines. Some of us were fortunate to have industrious parents who lived through the depressions of the twenties and thirties. They could see wars developing that would surely affect our generation. Long-term perspectives were altered by short-term, pressing economic problems. In navigation terms, we changed course often as conditions dictated, but always with the feeling that we would make it.

Times have changed. We now refer to the "American Dream." Does it still exist? Is it worth pursuing? I believe it is, but you need more than education alone. Inherited genes help, and you need persistence and a measure of the kind of good luck that comes to those who have best prepared. The dream is constantly changing. My opportunities did not even exist for my father's generation. This will be true for each generation to come.

My life embraced the entire world. I was not confined to one region. We knew that our generation was the first to be able to live like this, and appreciated the ease with which we could feel at home anywhere. However there was a penalty. No one can have it all. Because of our mobility, we could not participate in the community affairs of our own home area. Local politics, school events, and community activities were difficult to plan when you worked on a standby basis. We were familiar with the forest but not the trees.

Our Swiftly-Changing World

Strawberries in December? A passenger from "down under" traveled with us in 1947 to explore the idea. Was it practical to fly his strawberry crop to the U.S. (Spring and summer in Australia and New Zealand correspond to our fall and winter.) It seemed like a wild idea at the time, but it's routine today. Surely our diet and health are better as a result of having fresh foods available the year around.

Today complicated computer software is being created by brilliant East Indians in the most modest of environments. Television antennae are visible worldwide atop houses which do not even have running water. Tee shirts and blue jeans have become standard apparel everywhere. The U.S. has become the role model for the "third world," both a great opportunity and an enormous responsibility.

When people traveled by sailing ship, camel caravan, or on horseback, the earth was immense. A day's journey might cover ten to a hundred miles. But today, you can go anywhere in the world in one day. The globe has become very small.

Dulce Farmer,
my daughter.

Ellen Farmer,
my wife. (I'm so glad
she stopped by my
office that day to
say goodbye...and
stayed.

The loss of prairies, marshlands, and forests due to the population explosion in one lifetime should be a warning. If we continue, only a few generations will crowd us off our standard-of-living plateau. With too many people and too few resources, the U.S. will become as crowded as India or China. Of course, technology offers some hope, but the land mass in our temperate zone is limited.

Oil: I remember Dean Probert of the University of California School of Mines saying, in 1940, "Gentlemen, there are only oil reserves for another thirty years." This was before oil was developed in the Middle East and Alaska. Certainly, there will be more discoveries, but this resource has to end.

Air: Our layer of air is very thin, clean water is limited, agricultural land is not what it was fifty years ago. Wall Street and the government look at growth as a healthy sign. Yes—as long as we have room for expansion. Technology cannot manufacture space. The Japanese have been living with this reality for years.

Perhaps the biographies of my children and grandchildren will record that they have learned to be careful with their dwindling resources and shrinking space. Good luck!

An Ocean of Stars

History is rarely the story of individuals. Nameless ancient mariners crossed vast oceans, steering by the stars. The first flight navigators hung out of their planes to drop flares and make sightings, while their successors use highly sophisticated technology. Each advance has made the world smaller and accelerated the social, economic, and political changes that define our world. Today, space probes are navigating to distant galaxies, to be followed someday by manned space craft.

From the outrigger canoe to outer space, navigation has carried humanity as far as the imagination extends and farther still, constantly expanding the boundaries of the possible. I am proud to have been a small part of this giant push into the unexplored and uncharted, adding my contributions for the generations to follow.

Pan American Pacific Divsion World War II
Professional Navigators

(Subsequent activities, where known, follow in parenthesis.)

Albright, G.F.
Allen, Van (Navy Comm.)
Armbruster, J.
Baldocchi, Archie (investor)
Baldocchi, Don (cattle
  rancher)
Ballentyne, A.
Beeson, Joe (school
  principal)
Behrmann, Bob (Ph.D.,
  chemist)
Benson, V. (sporting
goods)
Bonner, H.
Bradish, Phil (lawyer)
Briggs, William
Brower, Joe (navigator for
  14 airlines, brother
  founded Sierra Club)
Buchanan, Bill
Cadwalader, C. (broker)
Campbell, Roy
Carson, Don
Casey, Walt (Las Vegas
  businessman)
Charman, A.L.
Conn, Bob (Pan Am Station
  Manager)
Cullom, F. (food broker)
Cummings, S.

Deans, A.M.
de Kramer, J. (artist,
  teacher)
Dickson, R.G. (F.A.A.)
Doige, R. (court reporter)
Donoghue, Jack
  (businessman)
Edgerly, R.
Erickson, V. (Los Angeles
  judge)
Fahnestock, B. (insurance)
Farmer, Mark
Fenn, Bob
Ferguson, Bob
Foley, Bill (Ph.D., chemist)
Francis, Al (Chief Nav.)
Gardner, O. (Los Angeles
  businessman)
George, Al
Goodwin, J. (teacher)
Grounds, Jack (travel agt.)
Harris, Ed (university prof.)
Harris, John (General
  Motors)
Harris, W.K.
Helmer, Jim (navigator,
  United Airlines)
Henderson, H. (Nevada
  businessman)
Hervey, T.E.
Hilbish, B. (Died in PB2Y3

INDEX

SHARE THE EXCITEMENT OF

FLIGHT TO ANYWHERE!

By Mark Farmer

To obtain additional copies, contact:

Bookmasters, Inc.
P.O. Box 388
Ashland, OH 44805
Tel: 800 247-6553

Or order on the internet from:

http://www.amazon.com

ISBN 0-9664832-0-0